The Handbook of Environmental Chemistry

Volume 4 Part E Air Pollution

O. Hutzinger
Editor-in-Chief

Springer-Verlag Berlin Heidelberg GmbH

Reactive Halogen Compounds in the Atmosphere

Volume Editors: P. Fabian and
Onkar N. Singh

With contributions by
T. Duafala, P. Fabian, M. J. Gillis,
J. Harnisch, M. A. K. Khalil,
A. McCulloch, P. M. Midgley, O. N. Singh,
R. Vogt

Springer

Environmental chemistry is a rather young and interdisciplinary field of science. Its aim is a complete description of the environment and of transformations occuring on a local or global scale. Environmental chemistry also gives an account of the impact of man's activities on the natural environment by describing observed changes.

"The Handbook of Environmental Chemistry" provides the compilation of today's knowledge. Contributions are written by leading experts with practical experience in their fields. The Handbook will grow with the increase in our scientific understandig and should provide a valuable source not only for scientists, but also for environmental managers and decision makers.

ISSN 1433-6855
ISBN 978-3-662-14745-0

Library of Congress Cataloging-in-Publication Data

The Natural environment and the biogeochemical cycles / with contributions by P. Craig ... [et al.].
v. <A–F> : ill. ; 25 cm. – (The Handbook of environmental chemistry : v. 1)
Includes bibliographical references and indexes.
ISBN 978-3-662-14745-0 ISBN 978-3-540-69690-2 (eBook)
DOI 10.1007/978-3-540-69690-2
1. Biogeochemical cycles. 2. Environmental chemistry. I. Craig, P. J., 1944– . II. Series.
QD31. H335 vol. 1 [QH344] 628.5 s [574.5'222]-DC20

Production Editor: ProduServ GmbH Verlagsservice, Berlin
Cover design: E. Kirchner, Springer-Verlag Berlin Heidelberg GmbH.
Typesetting: Fotosatz-Service Köhler GmbH, Würzburg
SPIN:10628761 52/3020 - 5 4 3 2 1 0 - Printed on acid-free paper

Advisory Board

Editor-in-Chief

Prof. Dr. em. Otto Hutzinger
Universität Bayreuth
Postfach 10 12 51
D-95440 Bayreuth, Germany
E-mail: otto.hutzinger@uni-bayreuth.de

Volume Editors

Prof. Dr. Peter Fabian
Lehrstuhl für Bioklimatologie
und Imissionsforschung
Ludwig-Maximilians-Universität München
Am Hochanger 13
D-85354 Freising-Weihenstephan, Germany
E-mail: fabian@met.forst.uni-muenchen.de

Prof. Dr. Onkar Nath Singh
Department of Applied Physics
Institute of Technology
Banaras Hindu University
221005 Varanasi, India
E-mail: onsingh@banaras.ernet.in

Preface

Environmental Chemistry is a relatively young science. Interest in this subject, however, is growing very rapidly and, although no agreement has been reached as yet about the exact content and limits of this interdisciplinary discipline, there appears to be increasing interest in seeing environmental topics which are based on chemistry embodied in this subject. One of the first objectives of Environmental Chemistry must be the study of the environment and of natural chemical processes which occur in the environment. A major purpose of this series on Environmental Chemistry, therefore, is to present a reasonably uniform view of various aspects of the chemistry of the environment and chemical reactions occurring in the environment.

The industrial activities of man have given a new dimension to Environmental Chemistry. We have now synthesized and described over five million chemical compounds and chemical industry produces about hundred and fifty million tons of synthetic chemicals annually. We ship billions of tons of oil per year and through mining operations and other geophysical modifications, large quantities of inorganic and organic materials are released from their natural deposits. Cities and metropolitan areas of up to 15 million inhabitants produce large quantities of waste in relatively small and confined areas. Much of the chemical products and waste products of modern society are released into the environment either during production, storage, transport, use or ultimate disposal. These released materials participate in natural cycles and reactions and frequently lead to interference and disturbance of natural systems.

Environmental Chemistry is concerned with reactions in the environment. It is about distribution and equilibria between environmental compartments. It is about reactions, pathways, thermodynamics and kinetics. An important purpose of this Handbook, is to aid understanding of the basic distribution and chemical reaction processes which occur in the environment.

Laws regulating toxic substances in various countries are designed to assess and control risk of chemicals to man and his environment. Science can contribute in two areas to this assessment; firstly in the area of toxicology and secondly in the area of chemical exposure. The available concentration ("environmental exposure concentration") depends on the fate of chemical compounds in the environment and thus their distribution and reaction behaviour in the environment. One very important contribution of Environmental Chemistry to the above mentioned toxic substances laws is to develop laboratory test methods, or mathematical correlations and models that predict the environmental fate of new chemical compounds. The third purpose of this Handbook is to help in the basic understanding and development of such test methods and models.

The last explicit purpose of the Handbook is to present, in concise form, the most important properties relating to environmental chemistry and hazard assessment for the most important series of chemical compounds.

At the moment three volumes of the Handbook are planned. Volume 1 deals with the natural environment and the biogeochemical cycles therein, including some background information such as energetics and ecology. Volume 2 is concerned with reactions and processes in the environment and deals with physical factors such as transport and adsorption, and chemical, photochemical and biochemical reactions in the environment, as well as some aspects of pharmacokinetics and metabolism within organisms. Volume 3 deals with anthropogenic compounds, their chemical backgrounds, production methods and information about their use, their environmental behaviour, analytical methodology and some important aspects of their toxic effects. The material for volume 1, 2 and 3 was each more than could easily be fitted into a single volume, and for this reason, as well as for the purpose of rapid publication of available manuscripts, all three volumes were divided in the parts A and B. Part A of all three volumes is now being published and the second part of each of these volumes should appear about six months thereafter. Publisher and editor hope to keep materials of the volumes one to three up to date and to extend coverage in the subject areas by publishing further parts in the future. Plans also exist for volumes dealing with different subject matter such as analysis, chemical technology and toxicology, and readers are encouraged to offer suggestions and advice as to future editions of "The Handbook of Environmental Chemistry".

Most chapters in the Handbook are written to a fairly advanced level and should be of interest to the graduate student and practising scientist. I also hope that the subject matter treated will be of interest to people outside chemistry and to scientists in industry as well as government and regulatory bodies. It would be very satisfying for me to see the books used as a basis for developing graduate courses in Environmental Chemistry.

Due to the breadth of the subject matter, it was not easy to edit this Handbook. Specialists had to be found in quite different areas of science who were willing to contribute a chapter within the prescribed schedule. It is with great satisfaction that I thank all 52 authors from 8 countries for their understanding and for devoting their time to this effort. Special thanks are due to Dr. F. Boschke of Springer for his advice and discussions throughout all stages of preparation of the Handbook. Mrs. A. Heinrich of Springer has significantly contributed to the technical development of the book through her conscientious and efficient work. Finally I like to thank my family, students and colleagues for being so patient with me during several critical phases of preparation for the Handbook, and to some colleagues and the secretaries for technical help.

I consider it a privilege to see my chosen subject grow. My interest in Environmental Chemistry dates back to my early college days in Vienna. I received significant impulses during my postdoctoral period at the University of California and my interest slowly developed during my time with the National Research

Council of Canada, before I could devote my full time of Environmental Chemistry, here in Amsterdam. I hope this Handbook may help deepen the interest of other scientists in this subject.

Amsterdam, May 1980 *O. Hutzinger*

Seventeen years have now passed since the appearance of the first volumes of the Handbook. Although the basic concept has remained the same some changes and adjustments were necessary.

Some years ago publishers and editor agreed to expand the Handbook by two new open-ended volume series: Air Pollution and Water Pollution. These broad topics could not be fitted easily into the headings of the first three volumes. All five volumes series are integrated through the choice of topics and by a system of cross referencing.

The outline of the Handbook is thus as follows:

1. The Natural Environment and the Biochemical Cycles,
2. Reactions and Processes,
3. Anthropogenic Compounds,
4. Air Pollution,
5. Water Pollution.

Rapid developments in Environmental Chemistry and the increasing breadth of the subject matter covered made it necessary to establish volume-editors. Each subject is not supervised by specialists in their respective fields.

A recent development is the 'Super Index', a subject index covering chapters of all published volumes, which will soon be available via the Springer Homepage http://www.springer.de or http://www.springer-ny.com or http://Link.springer.de.

With books in press and in preparation we have now published well over 30 volumes. Authors, volume-editors and editor-in-chief are rewarded by the broad acceptance of the 'Handbook' in the scientific community.

May 1997 *Otto Hutzinger*

Contents

Foreword

Reactive halogen compounds are important constituents of the natural atmosphere. The ocean, soils and the biosphere are natural sources of various bromine-; chlorine-; fluorine-; and iodine-containing substances whose reactivities vary within a large range, resulting in atmospheric lifetimes ranging from seconds to centuries. Man, through various activities, has also been releasing halogen compounds into the atmosphere. Some of these just added to the naturally emitted quantities and caused natural abundances to increase. Many of these man-made substances, however, were new species that had never before been part of the natural atmosphere. Depending on reactivities and emission rates, several substances have shown a substantial building up in their atmospheric abundances. Public concern arose when it became known that long-lived chlorofluorocarbons (CFCs) cause depletion of the ozone layer. Much of what is known today about halogen compounds in the atmosphere results from research initiated and/or intensified through this ozone depletion issue and its quantitative assessment. Based upon these findings, the Montreal Protocol and its later amendments, to phase out production and usage of substances that deplete the ozone layer, were adopted worldwide.

This handbook volume deals with halogen compounds, both natural and of anthropogenic origin, that are found in the atmosphere today. Sources, sinks and atmospheric budgets, global distributions, chemical pathways and the environmental impact of bromine, chlorine, fluorine and iodine species are described in the first four articles. Although reactive species in a strict sense comprise those with atmospheric lifetimes of, say, up to 10 years only, we have included longer-lived substances, too. While both natural and man-made halocarbons are covered in these four articles, articles five through eight are devoted to halocarbons entirely from industrial sources, their properties and applications, production, sales and emissions as well as international regulations related to the Montreal Protocol.

The wide scope of substances, their sources, applications and environmental impact, is reflected in the selection of authors, experts representing both academic as well as industrial views. Thus, a comprehensive volume covering most of the relevant references has been put together, with overlaps kept at minimum. Acknowledging the expertise and superb work of the authors, the volume editors hope that this volume will become a standard reference book for many years to come.

Weihenstephan and Varanasi, August 1998

Peter Fabian and
Onkar N. Singh

CHAPTER 1

Reactive Bromine Compounds

O. N. Singh[1] · P. Fabian[2]

[1] Department of Applied Physics, Institute of Technology, Banaras Hindu University, Varanasi- 221 005, India. *E-mail: onsing@banaras.ernet.in*

[2] University of Munich, Lehrstuhl für Bioklimatologie und Immissionsforschung, Am Hochanger 13, D-85354 Freising-Weihenstephan, Germany. *E-mail: fabian@met.forst.uni-muenchen.de*

Bromine, a minor constituent in the Earth's atmosphere – with its 50-fold higher efficiency of ozone destruction compared to chlorine – contributes significantly to the ozone hole formation and wintertime stratospheric ozone depletion over northern mid and high latitudes. In addition ozone episodes observed in the Arctic during polar sunrise are solely due to atmospheric bromine. CH_3Br, CH_2Br_2 and $CHBr_3$ are the major brominated gases in the atmosphere, of which CH_3Br being most abundant, contributes about 50% and CH_2Br_2 around 7 to 10% of the total organic stratospheric bromine. Bromocarbons with shorter lifetimes like $CHBr_3$, CH_2BrCl, $CHBr_2Cl$, $CHBrCl_2$ and CH_2BrI decompose before reaching the stratosphere, and are responsible for the ozone episodes. But for CH_3Br, which has also significant anthropogenic sources, all the aforementioned bromocarbons are mostly of marine origin. Halons (H-1211, H-1301, H-2402, H-1202) are solely anthropogenic and are far more stable. They decompose only after reaching the stratosphere. It is estimated that 39% of the stratospheric organic bromine ($\approx$ 7 pptv) loading is due to these halons. Increases are being still registered in the atmospheric abundance of halons in spite of production restrictions. Though extensively investigated, the existing knowledge with regard to the production and degradation of atmospheric bromine gases, is not commensurate with its importance.

Keywords: Bromine chemistry, bromocarbons, ozone depletion, ozone episode, ozone hole.

The Handbook of Environmental Chemistry Vol. 4 Part E
Reactive Halogen Compounds in the Atmosphere
(ed. by P. Fabian and O. N. Singh)

1 Introduction

Bromine is a minor constituent in the Earth's atmosphere. But for methyl bromide (CH_3Br) with an atmospheric mixing ratio of about 10 pptv (parts per trillion by volume), all other bromine bearing gases are only at a few pptv levels [1]. Around 7 pptv as bromine in the atmosphere comes from anthropogenic halons and 2 to 6 pptv from other naturally occurring bromine compounds [2]. Despite this low abundance, the bromine species play a vital role in our atmosphere.

The discovery of extraordinary ozone depletion in the Antarctic lower stratosphere [3, 4] and a little later in the Arctic lower troposphere [5, 6], popularly known as ozone hole and ozone episodes (also called surface ozone hole), have generated wide interest in the bromine chemistry, particularly in polar environments [7, 8]. Bromine per atom is about 50-fold more efficient than chlorine in converting ozone to oxygen. 20 to 25% of the austral springtime stratospheric ozone depletion over Antarctica, about one third of the wintertime stratospheric ozone depletion at northern mid and high latitudes, and almost 100% surface ozone destruction observed in the Arctic after polar sunrise are likely to be due to bromine. During the last two decades issues like sources, sinks, abundance and profiles, various chemical pathways and mechanisms etc. related to atmospheric bromine compounds and bromine have been vigorously addressed, and enormous amounts of monitoring and research work – experimental (observation and laboratory), theoretical and modelling – have been carried out and are still being continued.

Like most chlorofluorocarbons (CFCs), bromocarbons as such have high global warming potentials (GWP). But the recent comprehensive studies taking into account their share of stratospheric ozone destruction, predict a net cooling or considerably reduced radiative forcing of the Earth's surface due to further additions of bromocarbons. It may be pointed out that halons and methyl bromide with high GWP considered in the 1980s as double greenhouse gases, are now expected to slow down the pace of the anthropogenic greenhouse effect [9].

An attempt to review the current status of bromine species and related chemistry in our atmosphere is being made in this chapter.

2 Methyl Bromide

2.1 Introduction

Methyl bromide, the largest reservoir of atmospheric bromine, is likely to contribute at least 50% of the organic bromine budget to the Earth's atmosphere. It has both natural (ocean, wild fires) as well as anthropogenic (industrial production for agricultural and other kind of fumigation, biomass burning, gasoline additives) sources. Since halons, the other major bromine bearing substances, are being phased-out (under the regulations set out by the Montreal Protocol), methyl bromide has gained added importance. This is amply demonstrated by the amount of interest it has generated in the recent past – both in science and politics [1, 2, 9–35]. WMO, in its report no. 25 [12] states that if the global CH_3Br abundance was to be reduced by 10%, the ozone layer protection would be achieved approximately equivalent to an advance of the CFC phase out schedule by three years; and in the UNEP, 1992 report [16] it has been claimed that elimination of anthropogenic methyl bromide (estimated atmospheric abundance about 3 pptv) will provide an ozone layer protection comparable to that of an advance of the CFC and CCl_4 phase out schedule by 1.5–3 years. In accordance with the Montreal Protocol, its production for consumption in the developed countries has been capped at the 1991 level since the beginning of the year 1995 [1]. The US Environmental Protection Agency (US EPA) has already announced its complete phase-out by the year 2001 in the US [1] and global controls were established in 1995 to phase it out by the year 2010 [34]. Methyl bromide is also regulated domestically in a number of countries. The Netherlands phased out the use of methyl bromide for soil fumigation in 1992 because of ground water contamination concerns. Denmark and Nordic countries will ban agricultural use of methyl bromide by 1998, and Sweden is expected to follow a similar schedule. The European Union and Canada will cut agricultural use by 25% in 1998. A number of other countries are also contemplating regulatory action for methyl bromide use and production [34].

Since the atmospheric lifetime of CH_3Br is now estimated to be around one year (best estimate 0.7 year) [2, 35 a], cessation of its emissions should produce prompt results. The benefit of such action could be realised in a few years, whereas for the longer-lived CFCs, it would take until the year 2050 to reduce concentrations to what they were before the ozone hole developed. It will take centuries to reduce the atmospheric burden of the CFCs by a factor of 1,000; the atmospheric burden of anthropogenic CH_3Br could be reduced by that much within less than a decade [17].

Two extended and thorough reviews, under the auspices of UNEP [16] and WMO [1] respectively, have covered almost all aspects of atmospheric methyl bromide and have addressed to the uncertainties in its sources, sinks, budget, hemispheric distribution, lifetime and ozone depletion potential (ODP). At a cursory glance, these seem to present a complete and comprehensive description leaving no scope or need for further review. But the large amount of data generated and results published in the meantime and the investigations currently in progress, indicate that our understanding with regard to the various above mentioned uncertainties is rather sparse, and we find ourselves back to square number one.

Nevertheless, these two well written articles [1, 16] will be treated as the basis and will be extensively referred to. Much remains unclear about the origins and the impact of CH_3Br. Special sessions of two meetings: (i) Production and Fate of Atmospheric Organic Halogen in the Marine Environment, Spring Meeting of the AGU, Baltimore, Maryland 30 May-2 June 1995; and (ii) 1995 Methyl Bromide State of Science Workshop Monterey, California, 5–7 June 1995, were aimed to iron out some of the uncertainties [17]. Later, during 13–16 November 1995, at Hradec Kralove in the WMO Meeting "Consultation of Experts on Reactive Halogen Compounds and their Possible Effect on Ozone" these aspects were again discussed [18].

2.2 Measurements

Methyl bromide is an ubiquitous component of our lower atmosphere. Over the past two decades, sporadic measurements of surface abundance of methyl bromide have been carried out, and only a few measurements from the free troposphere or stratosphere, using balloon or aircraft platforms, are available [1, 19, 32]. A body of data from both hemispheres have been collected by several campaigns. The mean volume mixing ratios of methyl bromide in pptv, for both hemispheres are summarised in Table 1 [1, 10, 14, 22, 24–28]. In most cases, air samples were collected in pressurised stainless steel canisters and analysed after a period of several days, weeks or even months. Many measurements have been made with a technique involving sample preconcentration, gas chromatographic separation and electron capture detection (ECD) [1, 32]. Now it is being doubted that the levels of methyl bromide within these steel canisters can change over time, and also the results obtained from chromatographic analysis using ECD can depend upon the system configuration [10, 29]. These inferences are drawn on the basis of chromatographic analysis with mass spectrometric detection (GC–MS) of over 300 flask-pairs filled in parallel at seven different remote sites across the globe, and from the flasks filled during two oceanic cruises [10, 27, 29]. Levels of methyl bromide in stainless steel canisters have been observed to change up to a factor of two within two months time periods. These changes have been observed to varying degrees within a variety of different stainless steel sample containers and under a broad range of sample humidity [29].

There is a large range in the reported absolute volume mixing ratios – from 11 to 26 pptv in the NH and from 8 to 20 pptv in the Southern Hemisphere (SH) (Table 1, Fig. 1). Apparently, these differences are not due to a trend or seasonal fluctuations. These, therefore, could either be due to variations in the absolute

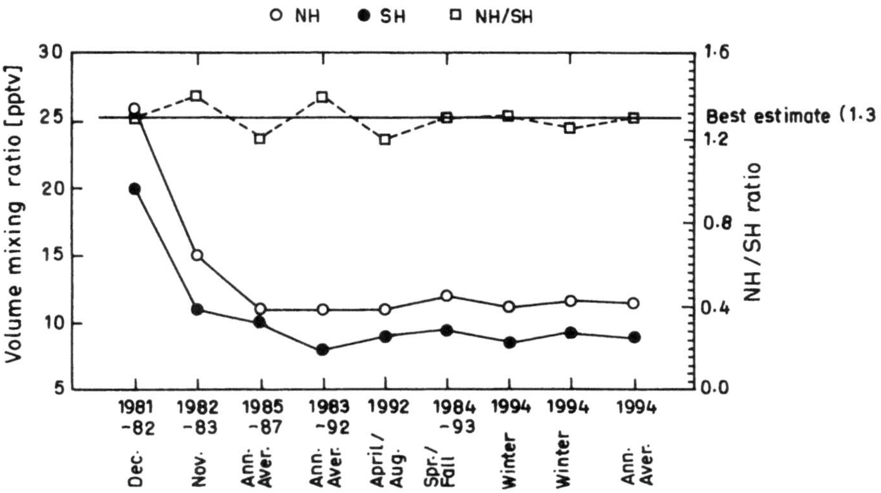

Fig. 1. Atmospheric concentrations of CH_3Br in the surface air of the Northern and the SHs, along with the Interhemispheric Ratio (NH/SH)

Table 1. Mean volume mixing ratios in pptv of CH_3Br in the surface air of the Northern and Southern Hemispheres

NH	SH	NH/SH	Year	Platform (Region)	Latitude Range	Ref. No.
26	20	1.3	1981–1982 (December)	Ship (Pacific)	40°N-32°S	[24]
15	11	1.4	1982–1983 (November)	Ship (Atlantic)	40°N-75°S	[25]
11	10	1.2	1985–1987 (Ann. Avg.)	Coastal (Pacific)	71°N-44°S	[22]
11	8	1.4	1983–1992 (Ann. Avg.)	Coastal (Pacific)	71°N-42°S	[14]
11	9	1.2	1992 (April/Aug.)	Coastal (Pacific)	90°N-45°S	[26]
12.0	9.5	1.3	1984–1993 (Spring/Fall)	Ship (Pacific, Atlantic)	60°N-90°S	[1]
11.2	8.6	1.31	1994 (Winter)	Ship (Pacific)	48°N-54°S	[10,27,28]
11.7	9.4	1.25	1994 (Winter)	Ship (Atlantic)	50°N-46°S	[10,27,28]
11.5	9.0	1.3	1994 (Ann. Avg.)	Ship (Pacific, Atlantic)	50°N-54°S	[10,27,28]

calibrations, to changes in volume mixing ratios taking place during the period of storage of the air or due to the configuration of GC-ECD equipment [10, 29]. Thus the key question is whether the observed abundance is truly representative of the background atmosphere in both hemispheres.

All investigations show that the abundance of methyl bromide is larger in the NH than that in the SH. The interhemispheric ratio (NH/SH) ranges from 1.2 to 1.4 (Table 1, Fig. 1). Perhaps the NH/SH ratio is more accurate compared to the measured absolute values because these ratios, if obtained from a single set of measurements, should not be sensitive to the absolute calibration [1, 16]. Some model studies predict the NH/SH ratio to be around 1.3 [30].

Taking into consideration the above mentioned uncertainties in the measurements and the absence of any distinct trend, we obtain, by simply averaging the reported absolute mixing ratios of the last ten years (1985 onwards) from Table 1, 11.3 ± 0.7 pptv and 9.1±1.1 pptv as average mixing ratios of methyl bromide in the NH and SH respectively. From these values, the global mean turns out to be 10.2 ± 0.9 pptv, which is about the currently accepted value of 10 pptv. The most acceptable interhemispheric ratio (NH/SH) is 1.3, as has been suggested in UNEP and WMO publications [1, 16]. The higher NH mixing ratios have been ascribed to the additional anthropogenic sources in the NH [1, 13, 15, 16]. This could also partly be due to larger CH_3Br sinks in the SH. Recently, a larger oceanic sink in the SH has been claimed [10] which takes into account about 0.2 pptv out of the inter-hemisphere difference of over 2 pptv.

Theoretical studies suggest that, due to the seasonal cycle of OH, CH_3Br should reveal a distinct seasonal trend. Further, this trend should be greater than the seasonal trends observed and predicted for methyl chloroform [1]. However, no clear cut seasonal cycle of CH_3Br has been detected so far [13, 14, 22]. Perhaps, this could be due to inadequate measurement precision and experimental artifacts, seasonal variation of methyl bromide sources and unidentified sinks [1].

The vertical distribution of CH_3Br has been found to decrease rather slowly with altitude in the troposphere [14, 31, 32]. At Hyderabad (17.5°N), Lal et al. [31] observed a sharp drop of its mixing ratio above the tropopause, i.e. 8.0 pptv at 20 km, 1.5 pptv at 23.6 km and 0.5 pptv at 25 km, whereas at Kiruna/Sweden (69°N), Fabian et al. [32] observed, due to a lower tropopause level, an even more rapid decrease with height, with mixing ratios at 20 km altitude already below the detection limit.

2.3 Sources

Four major sources – ocean, agriculture, biomass burning and gasoline additives; and two minor sources, structural purposes and industrial use of atmospheric methyl bromide – have been identified in the WMO report no. 37 [1]. A summary of the most likely contribution by each of these sources and their ranges is reproduced in Table 2. It was emphasised in this report that the existing state of knowledge with regard to atmospheric CH_3Br sources is imperfect. A detailed analysis and evaluation of each of these sources will be discussed in the following section.

2.3.1 *Oceanic Sources*

The oceans are the major natural reservoirs of bromine. They have generally been regarded as the principal natural source of CH_3Br. The only other natural source of CH_3Br is a minor one – the fraction of biomass burning emissions

Table 2. Emission of CH_3Br in thousand tonnes/year (best-estimates) [1, 10, 21, 27, 35, 38]

Source	Strength	Range	Anthropogenic	Natural
Ocean [a]	90	60–160	0	90
	-18	–	0	-18
	22.5	–	0	22.5
Agriculture	35	20–60	35	0
Biomass Burning	30	10–50	25	5
Gasoline Additives [a]	1	0.5–1.5	1	0
	15	9–22	15	0
Structural Purposes	4	4	4	0
Industrial Emissions	2	2	2	0
Totals [a]	162	97–278	67	95
	176	105–298	81	95
	54	19–100	67	-13
	68	27–120	81	-13
	95	60–141	67	28
	109	68–161	81	28

[a] Different measurements/estimates shown separately.

from wild fires. Singh et al. [24] observed, during December 1981, that the surface waters of the Pacific off North and South America [40 °N to 32 °S], were super-saturated with CH_3Br, with a mean supersaturation value of 250 %, clearly indicating these waters to be an important source. They extrapolated, on the basis of this data-set, an estimate of yearly global sea to air flux of 300 Gg/a (giga gram per annum). Ten years later Singh and Kanakidou [13] revised this figure to 40–80 Gg/a by correcting for large differences in calibration and by weighting the calculations according to the regional ocean productivity differences. Khalil et al. [14] obtained surface water supersaturation between 140 and 180 %, higher values for the open ocean compared to coastal sites, from their two pacific oceanic expeditions of 1983 and 1987 [67 °N to 50 °S]. Using these values they came up with a global oceanic flux of 35 Gg/a. For the purpose of comparison, by using their mean supersaturation of 160 % and a tropospheric mixing ratio of 11 pptv, along with the sea-air exchange and solubility coefficients used by Butler [11], a global oceanic flux of 45 Gg/a of CH_3Br was obtained [1].

From the two BLAST (Bromine Latitudinal Air-Sea Transect) data, the annual oceanic production and loss fluxes of CH_3Br were estimated to be 209 Gg and 227 Gg respectively [10, 27, 28]. This suggests an annual net flux of 18 Gg from the atmosphere to the ocean [10, 27], completely contrary to what has been accepted till recently. If correct, these results would overthrow the paradigm held so far that the ocean is the principal source of atmospheric methyl bromide leaving a lot of uncertainties in the atmospheric CH_3Br budget and making it difficult to explain the observed NH/SH ratio. Another production-based extrapolation made by Pilnis et al. [35] yields a positive global net flux of CH_3Br from the ocean to atmosphere, of 22.5 Gg/a. It may be pointed out that both these approaches suffer from the relatively limited geographical coverage of field measurements and from the limited understanding of the biological process involved in the production of CH_3Br in the water column. Currently there is hardly any basis for assessing the validity of the assumptions inherent in the two approaches [35].

Aquatic production of CH_3Br is most probably biological, yet so far no organism (or group of organisms) has been identified that can account for the oceanic emissions (about 60 Gg/a) required to support the observed saturation. Seaweed and ice algae do produce various halomethanes, but their global distribution cannot explain oceanic CH_3Br emissions. Planktonic diatoms typical of temperate to polar latitudes may produce CH_3Br, but not fast enough to support estimated oceanic emissions. Bacteria have not been found to be involved, although CH_3Br production apparently continues during decay of plant cultures [17].

2.3.2
Agricultural and Allied Usage

Methyl bromide is principally used as a fumigant, controlling a wide spectrum of pests, including pathogens, insects, rodents and nematodes. It has sufficient phytotoxicity to control many weeds and seeds in soil. It has features which make it a versatile and convenient material with a wide range of applications. In particular, it is quite penetrative, usually affective at low concentrations and leaves a residue which has generally been found acceptable. Its action is sufficiently fast and it airs rapidly enough from treated systems to cause relatively little disruption to commerce or crop production [33]. Due to its wide applications at low upfront costs, methyl bromide is important for the agricultural community.

Methyl bromide is normally supplied and transported in pressurised cylinders as a liquid, but at ambient temperature and pressure, CH_3Br is in the gaseous state. Methyl bromide is normally used directly from these cylinders or containers, but may sometimes be transferred to smaller units. Of the 1992

Table 3. Global sales of methyl bromide (tonnes) by use (China, India and former USSR not included) [33]

Year	Soil	Post Harvest	Structural	Residential/ Commercial	Chemical Intermediates	Total Sales
1984	30,408	9,001	1,285	881	3,997	45,572
1985	33,976	7,533	1,274	983	4,507	48,273
1986	36,090	8,332	1,030	999	4,004	50,455
1987	41,349	8,708	1,763	1,160	2,710	55,690
1988	45,131	8,028	1,910	1,737	3,804	60,610
1989	47,542	8,919	2,083	1,530	2,496	62,570
1990	51,306	8,411	1,740	1,494	3,693	66,644
1991	55,079	10,290	860	975	4,071	71,257
1992	57,407	9,564	902	1,062	2,648	71,583

Data source: Methyl Bromide Global Coalition (1994).

global sale of methyl bromide of 75.625 Gg, (75,625 tonnes), 3.2% was used as a feedstock for chemical synthesis. It is estimated that the remainder was used for soil treatment (76%), fumigation of durables (13%), fumigation of perishables (8.6%), and fumigation of structure and transportation (3%). The proportions of 1991, the base year, were similar to those for 1992. Global consumption excluding feedstock uses has increased by about 3.700 Gg/a (3700 tonnes per annum) since 1984 (Tables 3, 4, Fig. 2) [33].

Although methyl bromide clearly is a most useful substance for specific applications, there are a number of issues, not related to ozone depletion, which have led countries to impose restrictions on its use. Concerns include toxicity to humans and associated operator safety and public health, and residues. In some

Table 4. Sales of methyl bromide (tonnes) by region, including chemical feed-stock, but excluding China, India and former USSR [33]

Year	North America	South America	Europe	North Africa	Africa	Asia	Australasia	Total Sales
1984	19,659	1,389	11,364	183	1,595	10,687	704	45,572
1985	20,062	1,503	14,414	45	1,975	9,743	531	48,273
1986	20,410	1,774	13,870	380	2,205	11,278	538	50,455
1987	23,004	1,820	15,359	385	1,751	12,816	555	55,690
1988	24,848	2,058	17,478	277	1,582	13,555	812	60,610
1989	26,083	1,701	16,952	618	2,057	14,386	755	62,570
1990	28,101	1,621	19,119	432	1,838	14,605	928	66,644
1991	30,909	2,068	17,447	1,058	2,093	16,843	842	71,260
1992	29,466	2,300	18,521	1,363	1,697	16,944	1,294	71,585

Data source: Methyl Bromide Global Coalition (1994).

Note: Production tonnage for China, India and former USSR was estimated at 4,040 t for 1992.

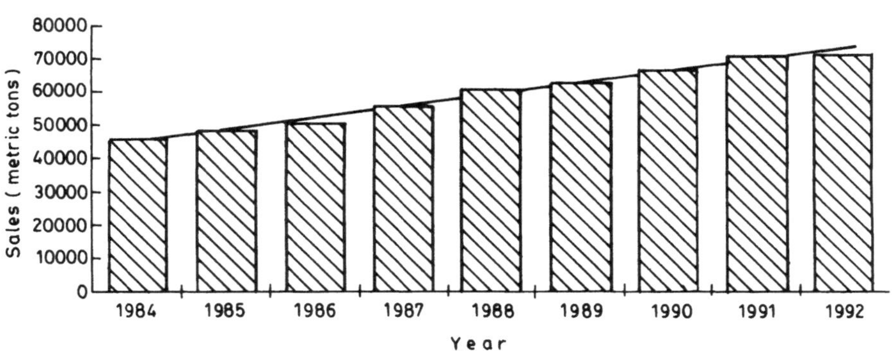

Fig. 2. Total global sales of methyl bromide (excluding China, India and former USSR) on an annual basis for the period 1984 to 1992 excluding feed stock uses [33]

countries, pollution of surface and ground water by methyl bromide and derived bromide ions are of concern, too [33].

Estimates of the fraction of methyl bromide released into the atmosphere vary widely. Emissions occur inadvertently through leakage and permeation during treatment, and unintentionally while venting at the end of treatments. The quantity of methyl bromide thereby emitted varies on an individual case basis as a result of the use pattern, the condition and nature of fumigated materials, the degree of seal of the enclosure and local environmental conditions. Methyl bromide is a reactive material: it is incorrect to equate production with emissions, as at least some of the methyl bromide applied is converted in use to non-volatile materials.

Under current usage patterns, the fractions of applied methyl bromide eventually emitted to the atmosphere, were estimated by the Methyl Bromide Technical Options Committee (MBTOC) to be 30 to 85%, 51 to 88%, 85 to 95% and 90 to 95% of applied dosage for soil, durables, perishable and structural treatments, respectively. These figures, weighted for proportion of use and particular treatments, correspond to a range of 46–81% overall emission from agricultural and related uses (34.00 to 59.00 Gg, based on 1992 sales data) [33]. There is active research into the development of recovery and recycling equipment for methyl bromide. It may be pointed out that most of the potential recovery and recycling systems are complex and may be expensive to install compared with the direct costs of fumigation facilities. Some systems would have high running costs associated with energy requirements. Many would require a level of technical competence to operate that would not normally be found at many fumigation facilities. If recovery is to be recognised as an acceptable method of reducing methyl bromide emissions to the atmosphere, it will be necessary to set specifications on aspects of fumigation, such as equipment efficiency and tolerable levels of emission [33].

2.3.3 *Biomass Burning*

The oceanic emissions, together with those from industrially manufactured methyl bromide fail to account for the total estimated atmospheric CH_3Br burden indicating the presence of some additional source. Taking cue from the presence of CH_3Cl in smoke plumes from biomass burning [36], it was speculated that biomass burning could as well be the missing source of CH_3Br [13, 14]. Recently, laboratory biomass combustion experiments showed that CH_3Br was emitted in various fuels tested. It was also detected in smoke plumes from wild fires in Savannas, Haparall and Boreal forests [21, 38].

In samples collected close to a Savanna grass fire in Kruger National Park [KNP] in South Africa, a large number of organic compounds were detected along with CO_2 and CO, [21]. The CH_3Br and CH_3Cl data from this fire are exhibited in Fig. 3 along with CO_2 and CO, [21]. Methyl chloride (CH_3Cl) shows a maximum during smouldering, whereas CH_3Br surprisingly shows two maxima, one during the flaming phase and another during smouldering. In slow burning

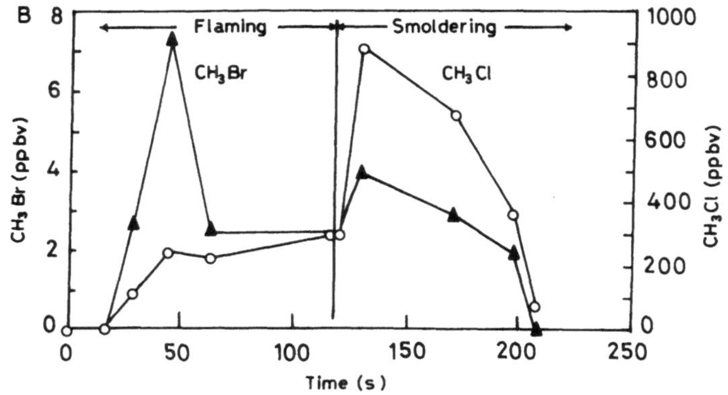

Fig. 3. Variation of CH_3Br and CH_3Cl concentrations with time in the smoke gas during the laboratory burn of South African Savanna grass fire [21]

fires, where there is little or no flame, CH_3Br and CH_3Cl both show one maximum only during smouldering [21].

Andreae et al. [38] from fuel samples, at KNP, obtained the average Cl and Br concentrations in the fuel to be 1260±310 and 7.0±1.6 mg kg^{-1} respectively. These values are somewhat higher than the concentrations of 650 mg Cl kg^{-1} and 4.0 mg Br kg^{-1}, deduced from the measurements by McKenzie et al. [39] in the Miombo Savanna, Zambia. The molar ratio Cl/Br in the KNP was found to be 406:1, a little higher than the ratio of 366:1 from Zambia [38,39]. The discrepancies between the two sets of values have been ascribed to the differences in the fuel mixtures. At KNP the fuel essentially consisted of grass, whereas large litter fractions with lower halogen content were abundant at the Miombo site. Nevertheless, Cl/Br ratios from both these small case Savanna sites were found to be larger than the Cl/Br ratio of about 100: 1 from forest vegetation of tropical as well as temperate regions. This explains the higher CH_3Cl/CH_3Br ratio in savanna smoke (120:1) as compared to the smoke from boreal forest fires in Siberia (53:1) and Finland (21:1) [21, 38]. During combustion, 90% of the halogen elements present in the fuel used were estimated to have actually been released to the atmosphere [38]. The CH_3Br concentrations in the smoke samples from the small and large fire plots were found to correlate well with the corresponding CH_3Cl and CO data [Fig. 4a and b).

Atmospheric emissions of methyl bromide could be estimated from the reference species CO, CO_2 or CH_3Cl. Based on CO and/or CH_3Cl, an emission estimate of about 7 Gg CH_3Br per annum from the Savanna fires has been arrived at. The CO_2 based flux was found to be relatively low. Although in the above estimation no formal assessment of the uncertainty was made so far, the overall uncertainty could be as large as a factor of 3 or more [38].

The pyrogenic source of CH_3Br has been assessed to range between 10–50 Gg/a [21], and 20–30 Gg/a [38], with the best estimate of 30 Gg/a. Thus the CH_3Br flux from biomass burning is of the same order of magnitude as that from the other major sources discussed before. Other estimates of this source

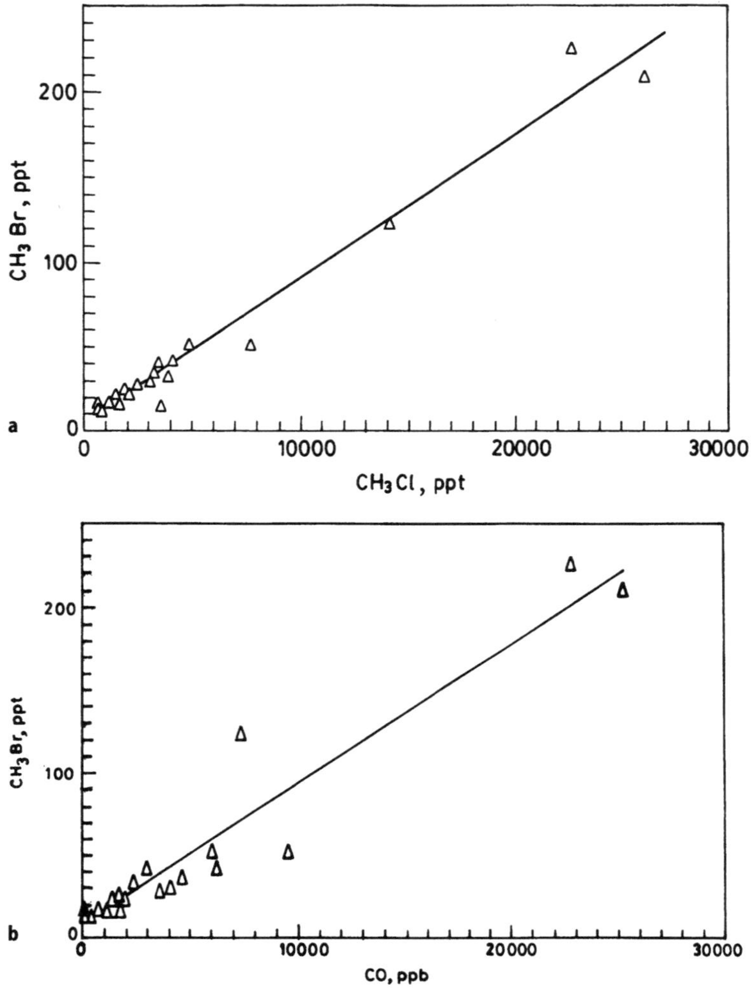

Fig. 4. Methyl bromide concentrations in samples from the Kruger Park fires plotted against **a** CH_3Cl and **b** CO [38]

were made by Penner et al. [40] who calculated a global inventory of CH_3Br and predicted annually averaged emissions from biomass burning to be 15 Gg/a, whereas Blake and Rowland, from their preliminary data, estimate a biomass burning source of 15–30 Gg/a [41].

2.3.4
Gasoline Additives and Other Sources

Methyl bromide is generated when leaded gasoline alloyed with ethylene dibromide is used as a fuel for internal combustion engines. It was estimated that between 1991 and 1992, 61 Gg of ethylene dibromide were used in such a fuel mixture world wide, out of which the US consumption was 24 Gg [1]. It may be added that in 1970 out of the global consumption of 170 ± 20 Gg [41 a], the US alone consumed 121 Gg of ethylene dibromide for this purpose, which was twice the global consumption in 1991–1992 [1]. This indicates that ethylene dibromide content in gasoline has been curtailed in the US; and the same has most likely been done in other industrialised countries. By 1995 world use of lead in gasoline has fallen to about 50 Gg [41 a]. Northern hemispheric consumption was 95% of the global in 1970 and remained 80% of the total in 1995 [41 a]. But on the other hand it is suspected that this reduction has been duly compensated by the increase in the use of CH_3Br for agricultural purposes [1].

In 1992, according to US EPA, about 100 billion gallons of leaded gasoline with ethylene bromide added, were consumed globally (2 billion gallons in the US), and the estimated amount of CH_3Br, ranging between 0.5 to 1.5 Gg, and 9 to 22 Gg, was emitted to the atmosphere (Table 2). There is a significant uncertainty in this evaluation, because reliable data from large parts of the world are not available; and also the estimates of the amout of CH_3Br emitted per Gg of leaded gasoline rangs from 15–28% to 0.1% [41 a].

About 4 Gg of CH_3Br is sprayed annually to fumigate structures like warehouses, historic buildings, food processing facilities, aircraft etc. Feasible alternatives are either being looked into or are already in use. As discussed before, CH_3Br is also used as an intermediate compound in manufacturing various chemicals. In the process, about 2.1 Gg/a of CH_3Br is inadvertently emitted to the atmosphere [1].

A great deal of uncertainty remains in the evaluation of the source strengths discussed. Special attention and more efforts are required for their better evaluation. If the oceans turn out to be a net sink, as has been pointed out earlier, re-evaluation of source strengths and search of possible additional sources may be required.

2.4 Sinks

Three sinks of atmospheric methyl bromide have been identified; (i) atmospheric removal, (ii) surface deposition, and (iii) oceanic removal. Till recently it was generally accepted that the only important sinks of atmospheric CH_3Br are reactions with tropospheric hydroxy radicals (OH) and irreversible loss to the ocean. But new investigations show soil degradation rates which, when extrapolated globally, represent a large additional natural sink [23, 18].

2.4.1 *Atmospheric Removal*

There are three possible removal mechanisms for CH_3Br in the atmosphere: (i) reaction with OH radical and other chemical species in the troposphere, (ii) precipitation in the troposphere, and (iii) transport and subsequent photolytic and other chemical removal in the stratosphere [1, 16]. The most significant of these is the removal by OH radicals in the troposphere. Methyl bromide does react with other species like NO_3 and Cl in the troposphere, but the removal rate through these is slow [1]. The tropospheric lifetime with respect to NO_3 is about 28 years, with respect to Cl even larger, of the order of 1000 years [1]. Removal by precipitation processes is thought to be unimportant, with an estimated lifetime of 2000 years [16]. A minor removal process is transport to the stratosphere followed by the reaction with the OH radical and photo-dissociation, with an estimated lifetime of between 30 and 40 years [1, 16].

Sufficiently consistent laboratory data for the rate constant of the gas phase OH reaction with CH_3Br have been generated [42, 43]. From the estimated lifetime of methyl chloroform (CH_3CCl_3) [44–46], which reacts with OH in the troposphere in a similar fashion, and the ratio of the rate constant of both OH reactions, the lifetime of CH_3Br with respect to OH is estimated as 1.7 years [42] and 2.1 years [43]. According to UNEP [16], WMO [1] and other studies [14], the overall life time of CH_3Br based on atmospheric removal alone is 2 years. However, Prinn et al. [47], estimated it as 1.67 years with a range of 1.52 to 1.86 years.

2.4.2
Surface Deposition

Methyl bromide, being a polar molecule and strong methylating agent, is expected to be subject to significant dry deposition [16]. Recently performed laboratory and field experiments [48, 49] indicate a robust and rapid microbial soil sink for atmospheric CH_3Br at current atmospheric mixing ratios. All surface soils tested consumed CH_3Br rapidly [48]. Temperate forest soils taken from 0–3 cm depth showed the fastest uptake rates reducing ambient CH_3Br levels to below the detection limit within 5 and 10 minutes. Activity declined with depth both in boreal and temperate forest soils. The uptake of CH_3Br seems to be an aerobic prokaryotic process. Consumption is stopped by autoclaving, and the temperature response is consistent with biological consumption. Maximum rates of uptake are between 25 and 37 °C. Activity is inhibited up to 98% by incubating soils under an anaerobic, nitrogen headspace. The bacteria which consume CH_3Br seem to thrive in warm and aerated soils [48].

The total estimated annual soil sink for CH_3Br is 42 ± 32 Gga (Table 5) [48]. Given a total atmospheric burden of CH_3Br of around 143 Gg (the estimated range of 100–200 Gg [1]) a soil loss term of 42 Gga^{-1} would correspond to a soil loss time of 3.4 years [2]. The main uncertainty in these findings is the degree to which the results obtained from a limited range of soils available for laboratory study and field work are representative of atmospheric CH_3Br uptake by soils throughout the world. Future measurements of more diverse sets of soil types and seasonality of soil sinks will be needed to clear up many of these uncertainties [48].

As yet there is no significant information available with regard to the methyl bromide depositions to the forest canopy, leaves and litter. A small diurnal variability in the ambient levels of methyl bromide observed at a rural site in Alabama has been ascribed to surface deposition at the forest canopy [50]. No such vari-

Table 5. Annual global soil uptake estimate for methyl bromide [48]

Biome	Area (m^2)	Active Season (d/a)	K_{obs} [a] (min^{-1})	Flux ($gm^{-2}d^{-1}$)	Total (10^9 g/a)
Tropical Forest and Savanna	22.5×10^{12}	365	0.0415 ± 0.0114	$(7.9 \pm 0.18) \times 10^{-7}$	6.5
Temperate Forest and Woodland/ Shrubland	20.0×10^{12}	240	0.245 ± 0.134	$(4.5 \pm 2.5) \times 10^{-6}$	21.7
Temperate Grassland	9.0×10^{12}	240	0.242 ± 0.079	$(4.5 \pm 1.5) \times 10^{-6}$	9.7
Boreal Forest	12.0×10^{12}	180	0.0894 ± 0.0002	$(7.9 \pm 1.8) \times 10^{-7}$	1.7
Cultivated Land	14.0×10^{12}	240	0.0805 ± 0.0421	$(7.9 \pm 1.8) \times 10^{-7}$	2.7
				Total	42.0

[a] K_{obs} is the observed first order rate constant for methyl bromide uptake into soils. For the soil types of temperate forest and woodland/shrubland and cultivated land, K_{obs} is the rate constant determined by field measurements. For the other soil types, K_{obs} is the rate constant of the surface soil layer as measured in laboratory incubations, and the stated error limits represent one standard deviation for all measured samples which varied between 2 and 5.

ability has been observed in the case of stable compounds such as CFCs and halons monitored simultaneously.

2.4.3
Oceanic Removal

Results obtained from two BLAST expeditions indicate that the ocean can act as a net sink [10, 27–29] for CH_3Br. Lobert et al. [10, 27, 28] found that coastal waters and regions of oceanic up-welling and current divergence were supersaturated in CH_3Br, whereas most of the open sea areas (representing 80% of the total oceanic surface) were unsaturated. They also observed that CH_3Br does not correlate with surface water CH_3Cl or CH_3I as believed earlier [13, 24], and hence surface water CH_3Br cannot be predicted from the measurements of CH_3Cl and CH_3I. Unlike methyl bromide, both methyl chloride and methyl iodide were largely supersaturated everywhere in the ocean [10, 27, 28].

The partial pressure profile of methyl bromide in surface waters was observed by Lobert et al. [10, 27, 28] to be distinctly different from the atmospheric profile and was generally lower than that in the air, with exceptions in coastal, coastal influenced and up-welling regions. Using these partial pressures (PP) of CH_3Br in air and water as an indicator for the departure of the dissolved amount from equilibrium, they estimated the saturation anomaly. For evaluating the net saturation anomaly (NSA), they also took into account the influence of physical effects like seasonal temperature variation of water, advection and mixing. Positive or negative NSA was taken as an index of surface emission or loss of CH_3Br in sea water. While estimating the total oceanic production of methyl bromide from its observed net flux, hydrolysis loss, nucleophilic chloride substitution [51, 52] and downward removal [11] of the compound were duly accounted for [28].

Pilinis et al. [35], also using data from shipboard measurements of saturation anomalies, extrapolated the ocean/atmospheric flux of CH_3Br on the basis of steady state production rate of CH_3Br in the water column, allowing the regional and seasonal variation in temperature to affect the saturation. Their simulation yielded the ocean to be a strong net source of CH_3Br to the atmosphere. The difference between this estimate and that of Lobert et al. [10] was mainly assigned to the slow degradation rates in water at higher latitudes. In order to switch the ocean from source to sink, the applied production rate of Pilinis et al. [35] needs to be reduced by over 35%. This amply demonstrates how sensitive current estimates of oceanic fluxes are to assumptions with regard to CH_3Br production and destruction in the water column. The saturation based extrapolation yields a negative global net flux of CH_3Br from the ocean to the atmosphere [10], while production based extrapolations yield a global net positive flux [35].

King et al. [51] determined the chemical loss rate of CH_3Br in sea water. They estimated the global mean value for the chemical loss rate of CH_3Br in sea water to be 8.8% per day. The loss rate varies by more than two orders of magnitude between polar and equatorial waters. The global loss rate suggests that oceans are operating as a buffer towards changes in atmospheric CH_3Br. Butler's model [11] predicts that the ocean will buffer the atmosphere against changes in the atmospheric concentrations induced by the reduction in agricultural utilisation, but King et al. [51] conclude that the oceanic chemical destruction rates vary, from slow rates in cold waters, where the oceans would not provide an effective buffer, to rapid rates in warm waters, where the oceans are near their upper limits in terms of buffering effectiveness. Further, the oceans as a whole have only about 1/3 of their maximum theoretical buffering capacity of Butler's estimate. If other destruction mechanisms, such as biological uptake, are found to occur in the oceans, this value may increase. King et al. [51] have observed that degradation of CH_3Br in unfiltered sea water taken from the southern Florida coast is up to twice as rapid as the abiotic removal rate.

The major abiotic oceanic degradation pathways of CH_3Br are by hydrolysis and by exchange with chloride ions to yield CH_3Cl. King et al. [51] give a single first-order "sea water" rate constant based on measurements from 10 to 35 °C, but they do not separate out the two mechanisms or the specific rate dependence on chloride ions. Elliot and Rowland [52] did report separate rate constants for hydrolysis and chloride exchange of CH_3Br, but their exchange measurements were limited to two temperatures 0 and 22 °C. Later they reported [53] values for the hydrolysis rate constants at 9 and 22 °C, measured in distilled water, 0.5 molar NaCl, and sea water. Butler [11] and Yvon and Butler [2, 35a] have incorporated the major known pathways into a model that allows determination of the critical elements in establishing a global balance of methyl bromide. Jeffers and Wolfe [54] carried out experiments to establish the Arrhenius behaviour of the hydrolysis and chloride exchange reactions and showed that the exchange reaction is first-order in both CH_3Br and chloride ion. They experimentally covered a sufficiently wide temperature range to allow for precise extrapolations of rate constants to all significant ocean water temperatures. They observed that a change of sea water temperature from 15 to 30 °C, both well within the natural range of ocean temperatures, results in a change in the CH_3Br degradation rate constant by a factor of nearly 10, and a change in aquatic half-life from 16 days to under 2 days. Jeffers and Wolfe [54], in a series of experiments found that green plants can absorb or metabolise CH_3Br from air. Oremland et al. [55] observed indirect biodegradation of CH_3Br in anaerobic salt marsh sediments due to sulphide attack with the sulphide produced by methanogenic bacteria.

The abiotic degradation parameters should allow accurate integration of CH_3Br loss over the thermal and salinity variation of the Earth's water surfaces. All these additional tropospheric sinks (including the soil degradation), if properly characterised, would allow an accurate and fair evaluation of the overall influence of human production and use of CH_3Br. The earlier best estimate of the lifetime of methyl bromide with respect to oceanic loss was 3.7 years with a possible range of 1.4 to 14 years [11]; this has now been modified to 2.7 and 1.9 years with full possible range of 2.4 to 6.5 and 1.1 to 3.9 years respectively [2, 35 a].

2.5 Lifetime and Ozone Depletion Potential

From the lifetimes of methyl bromide with respect to different loss processes – atmospheric removal, $\tau_a = 1.7$ years, surface deposition, $\tau_s = 3.4$ years, and oceanic removal, $\tau_o = 1.9$ years – the total life time (τ) of methyl bromide can be evaluated according to [56]:

$$1/\tau = 1/\tau_a + 1/\tau_s + 1/\tau_o$$

Presently the best estimate for τ of methyl bromide is 0.7 years with a full range of 0.6 to 0.9 years [2, 35 a]. This value is considerably smaller than the estimates of 2 years and 1.3 years in the UNEP [16] and WMO reports [1] respectively. Since the uncertainty in the lifetime is reflected in the ozone depletion potential (ODP), an important parameter both for science and politics, a good amount of effort has been and is being made for the accurate determination of the total lifetime of methyl bromide.

The ODP of a substance is a measure of the amount of ozone destruction resulting from the release of a certain mass of the substance relative to that resulting from the same mass of CFC-11 released to the atmosphere [1, 12–16, 57–59]. Its primary purpose is for comparison of relative impacts of different gases on ozone. ODP is a quantity that cannot be measured directly in the laboratory. It can only be evaluated by using empirical or numerical models. The ODP of methyl bromide depends on the efficiency of bromine chemistry in removing ozone in the stratosphere and on the amount of bromine released in the stratosphere from methyl bromide. Steady state ODPs represent the cumulative effect on ozone over an infinite time scale, whereas time dependent ODPs describe the temporal evolution of this ozone impact over specific time horizons [1, 12, 60]. Models used to evaluate ODPs now include better representations of mid-latitude and polar vortex heterogeneous chemistry processes than those used earlier [1].

The rate limiting steps for ozone removal are given by the reactions of BrO with ClO, O, HO_2 and BrO [1, 57–60]. Thus, the ability to calculate the ODPs for the bromine species depends on the ability to calculate the precise partitioning of the radical species, i.e., what portion is in the form of BrO as well as those of ClO, O and HO_2. Using the AER-2D chemistry transport model, concentrations of BrO and HBr were calculated [47] for this purpose and then compared with the available field data of BrO [61–66] and HBr [67].

Using 0.8 yr as the lifetime of CH_3Br, its ODP is estimated as 0.36 with a full range of 0.29 to 0.57 [2], which will be still smaller if the lifetime used is 0.7 yr [35a]. Taking the lifetime to be 0.9 years and the updated bromine efficiency of 46, the ODP value of CH_3Br is evaluated to be 0.4 [48, 68, 69], a value considerably smaller than the values published in the UNEP and WMO reports [1, 16].

3 Other Bromocarbons in the Earth's Atmosphere

3.1 *Introduction*

Bromocarbons having shorter lifetimes compared to methyl bromide, like bromoform or tribromomethane ($CHBr_3$), methylene bromide or dibromomethane (CH_2Br_2), bromo-chloromethane (CH_2BrCl), dibromochloromethane ($CHBr_2Cl$), bromodichloro-methane ($CHBrCl_2$), bromoethane (C_2H_5Br) and bromoiodomethane (CH_2BrI) are also important bromine gases in the atmosphere. They are mainly responsible for the sensational ozone episodes (discussed in detail under 3.3) occurring during polar sunrise in the Arctic lower troposphere. Except for dibromomethane which is estimated to contribute about 7 to 10% of organobromine [19, 69, 70] their contribution to stratospheric bromine budget is negligible. CH_2BrCl, $CHBr_2Cl$, $CHBrCl_2$ and CH_2BrI are decomposed before reaching the stratosphere and seem to be of lesser importance even for the tropospheric bromine budget due to their lower abundance. The total abundance of these compounds is estimated to be less than 1 pptv [18].

Only sporadic measurements of these bromocarbons are available. Almost all the measurements carried out thus far are from polar regions. In the 1980s, the detection and measurements of these species have been made by Rasmussen and Khalil [71], Penkett et al. [72] Khalil and Rasmussen [73] Berg et al. [74], Cicerone et al. [22], Class and Ballschmiter [75]. Since the discovery of ozone episodes in the late 1980s, the interest in the bromine chemistry in the lower troposphere of the Arctic has gained momentum. A few comprehensive measurements of some of these bromine bearing gases have been carried out under the auspices of Polar Sunrise Experiments (PSE-1988, PSE-1992 and PSE-1995) [6, 76, 76a]. In addition, some other measurements have also been published [77–79]. In the Arctic, higher concentrations of $CHBr_3$ and $CHClBr_2$ have been recorded in winter and spring as compared to summer, whereas no significant seasonal variation has been observed in the abundance of CH_2ClBr (Fig. 5a–c).

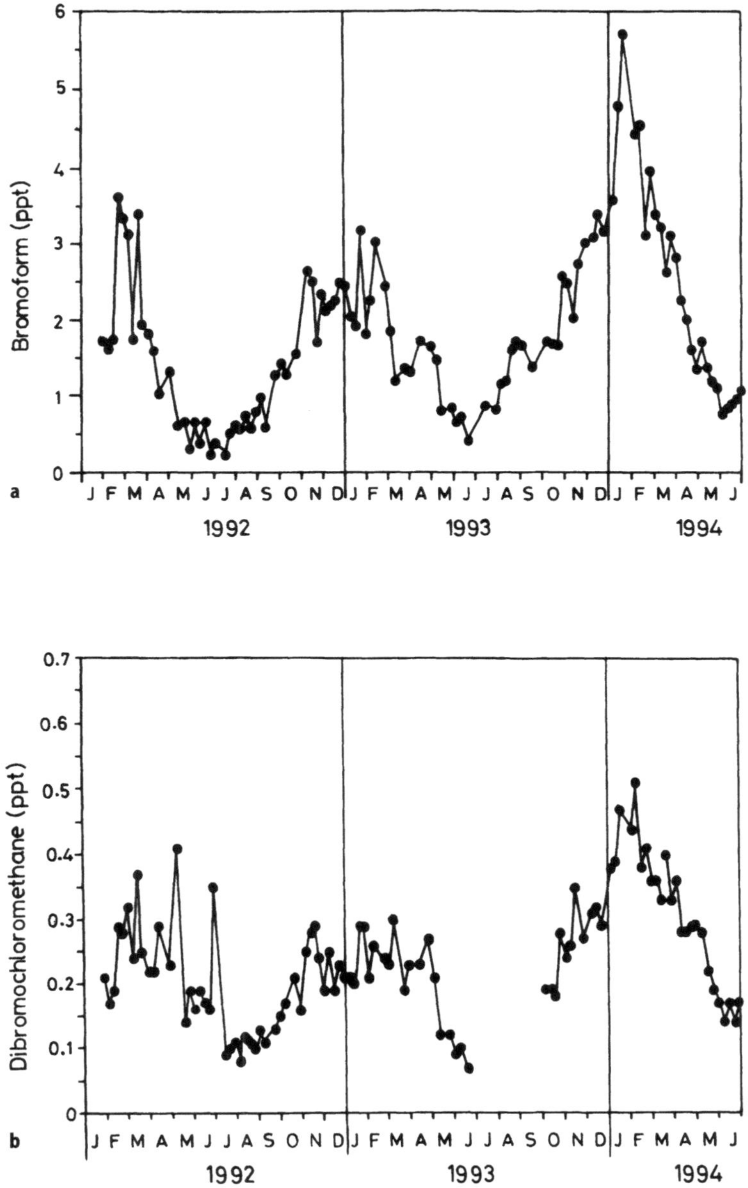

Fig. 5. a The seasonal variation of weekly average bromoform concentrations at the ground level at Alert [79]. **b** The seasonal variation of weekly average dibromochloromethane concentrations at the ground level at Alert [79]

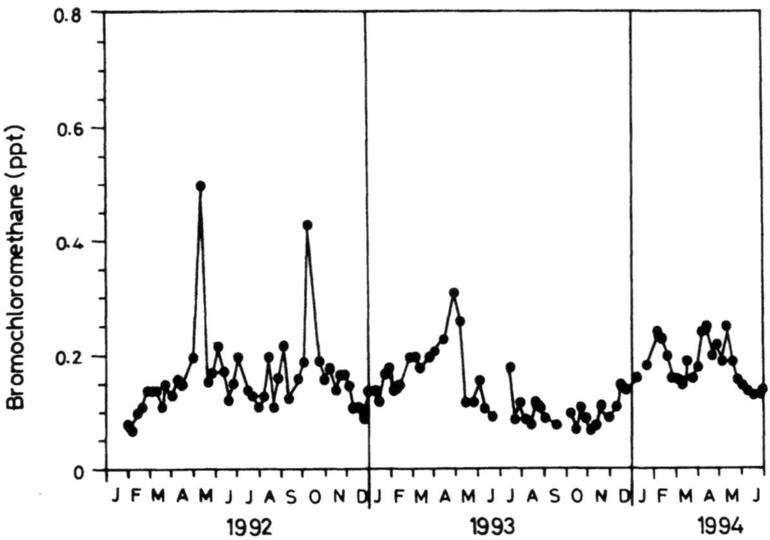

Fig. 5c. The seasonal variation of weekly average bromochloromethane concentrations at the ground level at Alert [79]

Maximum (winter) and minimum (summer) weekly averages measured at Alert Northwest Territories, Canada (82.5 °N, 62.3 °W) between January 1992 and June 1994 in pptv are 3.0 and 0.84, 0.30 and 0.14, and 0.15 and 0.14, respectively for $CHBr_3$, $CHClBr_2$ and CH_2ClBr, Fig. 5 [79].

3.2 Sources and Measurements

Most of these bromine compounds are of biogenic origin, although their production may involve anthropogenic substances [76]. These are produced by various types of oceanic water and ice algae and phytoplankton [78–85]. $CHBr_3$, $CHBr_2Cl$ and CH_2Br_2 are consistently released by algae to the sea water at rates ranging from ng to μg of each compound per gram of dry algae per day. Biogenic release of $CHBr_3$, particularly from ice algae, is considered to be its significant source [78, 81–83]. In the Arctic, sea weeds presumably constitute a small but potential source of organohalogens [78, 84]. A detailed quantitative estimation of the rate of release of these organobromine compounds by different algal species, particularly from sea ice microalgae (both in the Arctic and Antarctic) has been made [78–85]. Other probable sources of these organobromines are yet to be properly investigated. Some new data with regard to $CHBr_3$ and CH_2Br_2 measurements along with CH_3Br are expected from the BLAST-1994, campaign [28].

Since observational data records of sufficient frequency, duration and extent are not available, it is not possible to determine the global trends of these bromine compounds. CH_2Br_2 and $CHBr_3$ exhibit large variability in the troposphere partly due to their shorter lifetime but also due to the variability in their source strength distributions. Measured concentrations at ground level vary around 1–3 pptv and 1–40 pptv for CH_2Br_2 and $CHBr_3$ respectively [18, 79]. The average mixing ratios of CH_2Br_2, $CHBr_3$ and C_2H_5Br have been estimated to be between 2 and 3 pptv, of CH_2BrCl between 1 to 2 pptv, and of $CHBr_2Cl$ and $CHBrCl_2$ about 1pptv (Table 6) [12]. According to recent estimates, the abundance of the CH_2BrCl, $CHBr_2Cl$, $CHBrCl_2$ and CH_2IBr are below the 1 pptv level [18].

Under the auspicious of PSE-1988, Bottenheim et al. [76] obtained the abundance of bromoform between 1 and 10 pptv in air samples collected at Alert during March/April. These air samples were independently analysed for $CHBr_3$ at three centres, and three sets of data were generated. The agreement in the volume mixing ratios from different centres were poor. Later in 1990 the abundance of $CHBr_3$, CH_2Br_2, $CHBr_2Cl$, $CHBrCl_2$ and CH_2BrCl were measured by Sturges and Cota from air samples collected at Barrow, Alaska (71.3 °N, 156.6 °W) and Resolute Canada (74.7 °N, 94.9 °W). They found atmospheric bromine levels dominated by bromoform (0 to 4 pptv), followed by dibromomethane (Fig. 6a [78]). They also measured the release of bromoform to sea water in ice algal incubation and came up with the estimate that 1.9 to 19% of the measured bromoform production, upon reaching the surface, would account for its atmospheric burden. But actual measured fluxes of bromoform through holes in the ice fell short of the required flux rate. The observed high concentrations of bromoform in surface sea ice and snow suggested an additional contribution to atmospheric levels directly from the snow surface. At Resolute, bromoform was measured during spring of both 1991 and 1992 (Fig. 6b) [78].

Table 6. Atmospheric mixing ratios of some bromocarbons [12, 18]

Compound	Melting point (°C)	Boiling point (°C)	Volume mixing Ratio pptv
Methyl Bromide (CH_3Br)	- 97.7	3.56	10
Methylene Bromide (CH_2Br_2)	- 52.7	96.9	2-3
Bromoform ($CHBr_3$)	7.7	149.5	2-3
Bromochloromethane (CH_2BrCl)	- 88	68.1	1-2
			< 1
Dibromochloromethane ($CHBr_2Cl$)	- 22	119–120	1
			< 1
Bromodichloromethane ($CHBrCl_2$)	- 57.1	90.1	1
			< 1
Bromoiodomethane (CH_2IBr)	–	138–141	< 1
Ethylidene bromide ($C_2H_4Br_2$)	9.9	131.4	1
Ethylbromide (C_2H_5Br)	- 119.3	38.4	2-3
Halons:			
H-1211 ($CBrClF_2$)	- 160.5	- 4.01	2-3
H-1301 ($CBrF_3$)	- 168	- 58 to -57	1-3
H-2402 ($C_2Br_2F_4$)	- 110.5	47.3	0.2-0.5
H-1202 (CF_2Br_2)	- 142 to - 141	22–23	0.03

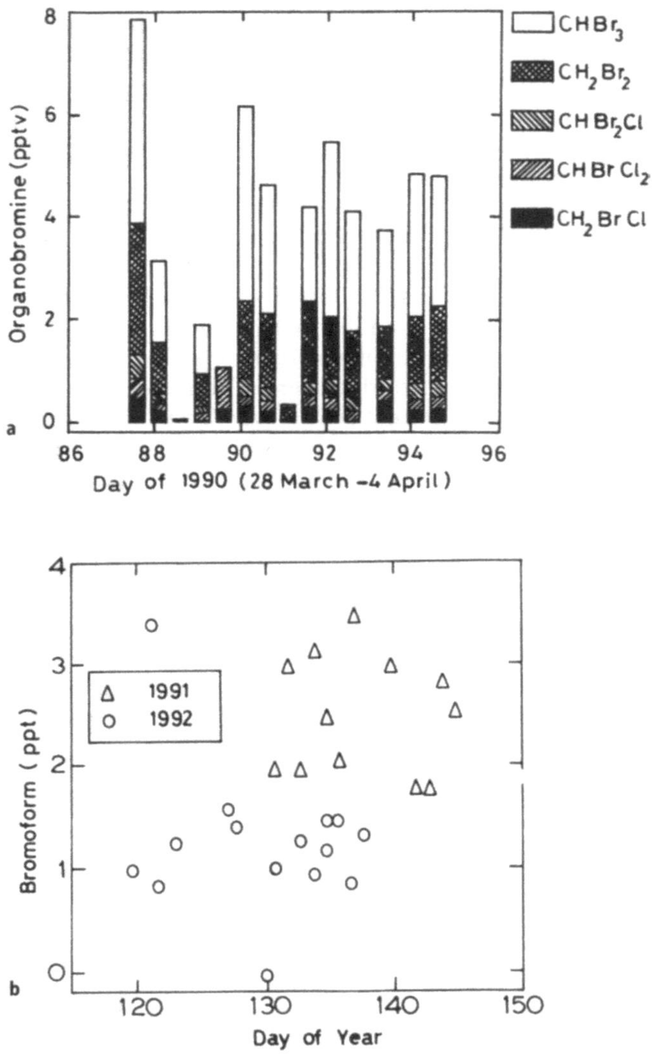

Fig. 6. a Atmospheric concentrations of organobromine compounds measured at Barrow, Alaska during the spring of 1990 [78]. **b** Summary of atmospheric bromoform measurements at Resolute, during the spring of 1991 and 1992 [78]

Under the auspices of PSE-1992, the volume mixing ratios of $CHBr_3$, CH_2Br_2, $CHBr_2Cl$ and $CHCl_2Br$ were obtained by several groups from ground based and low altitude aircraft measurements using several type of filters [86–89]. Continuous monitoring was also carried out for a few weeks to a few months with the intention of observing correlations between the volume mixing ratios of these organobromine compounds with surface ozone concentrations during ozone events.

Yokouchi et al. [86] measured the volume mixing ratio of $CHBr_3$ in the range of 2.0 to 3.7 pptv, with an average of 2.6 pptv, and CH_2Br_2 and $CHBr_2Cl$ in the range of 0.5 to 1 pptv and 0.1 to 0.5 pptv respectively. Hopper et al. [87], in April 1992, measured $CHBr_3$ mixing ratios to be greater than 7 pptv at SWAN (83.9 °N, 63.1 °W, in the Arctic ocean), which was more than twice the value obtained at the same time at Alert. Similarly, the abundance of CH_2Br_2 was measured as 1.89 pptv, much higher than that at Alert (0.4 pptv). Airborne measurements by Leaitch et al. [88] yielded volume mixing ratios in pptv of $CHBr_3$, $CHClBr_2$, CH_2Br_2 and $CHCl_2Br$ to be 0.8 to 2.5, 0.18 to 0.36, 0.7 to 1.7 and 0.4 to 0.73 respectively. Yokounchi et al. [79] recently reported weekly average concentrations of $CHBr_3$, $CHClBr_2$ and CH_2ClBr from measurements at Alert carried out from January 1992 to June 1994. A comparison of measured concentrations from different sites and times of bromoform and dibromochloromethane is shown in Table 7 and Fig. 7 (a, b) [79]. The ambient concentrations of $CHBr_3$ and CH_2Br_2 over the western Pacific during 1991/92 were measured in the range of 0.13–2.9 and 0.14–1.58 pptv, respectively. But the concentrations of $CHBr_2$ as high as 7 and 3.5 pptv were observed in the samples collected respectively in the harbours of Singapore and Pinang during Jan–March 1994 [79 a].

Particle and gas phase inorganic bromine, total organic bromine and also $CHBr_3$, CH_2Br_2, $CHCl_2Br$ and $CHClBr_2$ were individually measured at Alert continuously during January 18 to April 21, 1992 by Li et al. [89]. The total organic bromine was determined from the charcoal traps and instrumental neutron activation analysis. Several individual organic species of bromine

Table 7. Comparison of measured concentrations of $CHBr_3$ and $CHBr_2Cl$ in the Arctic atmosphere

Compound	Site	Time	Concentration (pptv)	Ref.
Bromoform	Mould Bay	1983 spring	5–40	[74]
($CHBr_3$)	Ny-Ålesund	1992 September	<0.02–1.19	[77]
	Ny-Ålesund	1983 spring	3–40	[74]
	Barrow	1983 spring	2–46	[74]
	Barrow	1984–1987	2–11	[22]
	Alert	1988 spring	1–10	[76]
	Alert	1992–1994	0.25–5.7	[79]
	Western Pacific	Jan–Feb 1991	0.28–2.9	[79 a]
	Western Pacific	Sep–Oct 1992	0.13–1.91	[79 a]
	East and Southeast Asian Seas	Jan–March 1994	0.31–7.1	[79 a]
Dibromochloromethane	Ny-Ålesund	1992 September	<0.01–1.00	[77]
($CHBr_2Cl$)	Alert	1988 spring	0.2–0.5	[76]
	Alert	1992–1994	0.07–0.5	[79]
Dibromomethane	Western Pacific	Jan–Feb 1991	0.47–1.36	[79 a]
(CH_2Br_2)	Western Pacific	Sep–Oct 1992	0.14–1.58	[79 a]
	East and Southeast	Jan–March	0.38–1.42	[79 a]

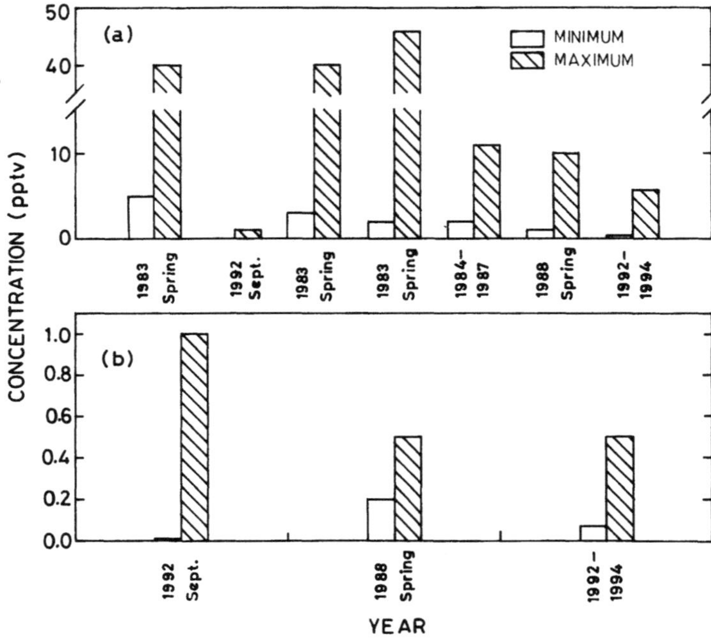

Fig. 7. Comparison of measured concentrations of **a** bromoform ($CHBr_3$) **b** dibromochloromethane ($CHBr_2Cl$)

were detected using two different methods, one with GC/MS using a Tenax trap and the other with charcoal trap and GC-ECD detector. The results are shown in Fig. 8 [89].

3.3 Arctic Surface Ozone Hole

During polar nights, photochemistry is non-existent in the Arctic atmosphere due to the absence of light for months. Consequently, photochemically reactive species accumulate and are abundant at the time of polar sunrise. Thus, when sunlight returns, the Arctic atmosphere undergoes a change caused by photochemically induced reactions. One of such reactions is the conversion of marine halogens, particularly bromine, to filterable halogens, supposedly at times leading to a complete destruction of surface ozone from its ambient concentrations of between 40 and 45 ppbv to undetectable limits [5, 76, 78, 86–100]. Field measurements at Alert [100] have shown that surface ozone concentrations were nearly constant (25–30 ppbv) during the polar dark period, January to mid February. By early March they had reached 40 ppbv and by mid-April about 45 ppbv. Three major ozone depletion events were observed in April 1992, with ozone less than 0.5 ppbv, the instrumental detection limit. Inorganic and organic gaseous bromide and particulate bromide regularly peaked with these

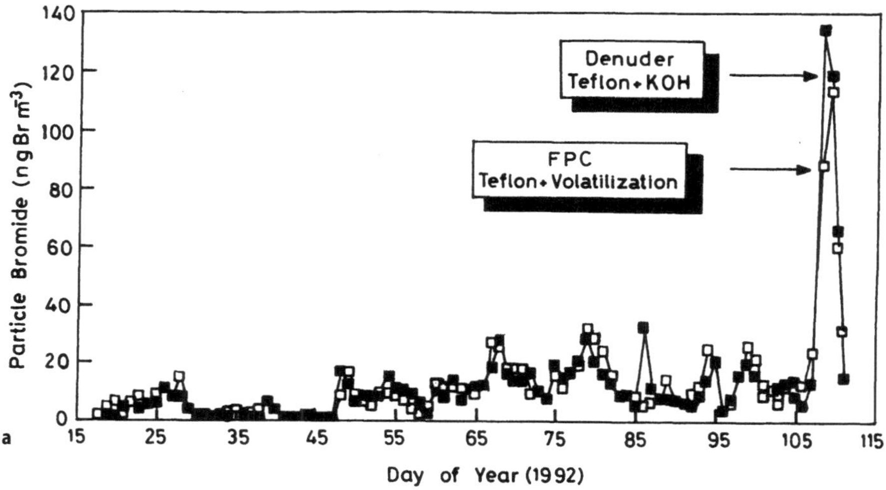

a

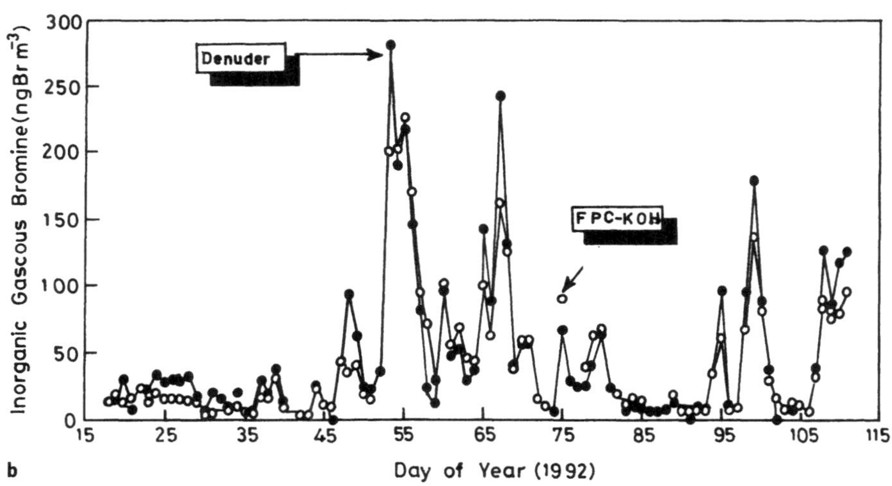

b

Fig. 8. **a** Particle bromine during the Polar Sunrise 1992 Experiment at Alert [89]; **b** the time series of air concentrations of inorganic gaseous bromine (InorgBr) at Alert determined using two different sampling/analysis methods [89]

major and minor ozone depletion events [92]. These remarkable episodes of ozone destruction in the Arctic lower troposphere were observed year after year at the time of polar sunrise. The strong anti-correlation with filterable bromine, to some extent with filterable iodine as well, another substance of mainly oceanic origin [38] have led to several proposals towards mechanisms [5, 76, 95, 98, 99] and related measurements [6].

Aerosal chemical mass-size distribution measurements were carried out during PSE-1992 primarily to explain the chemical mechanism responsible for

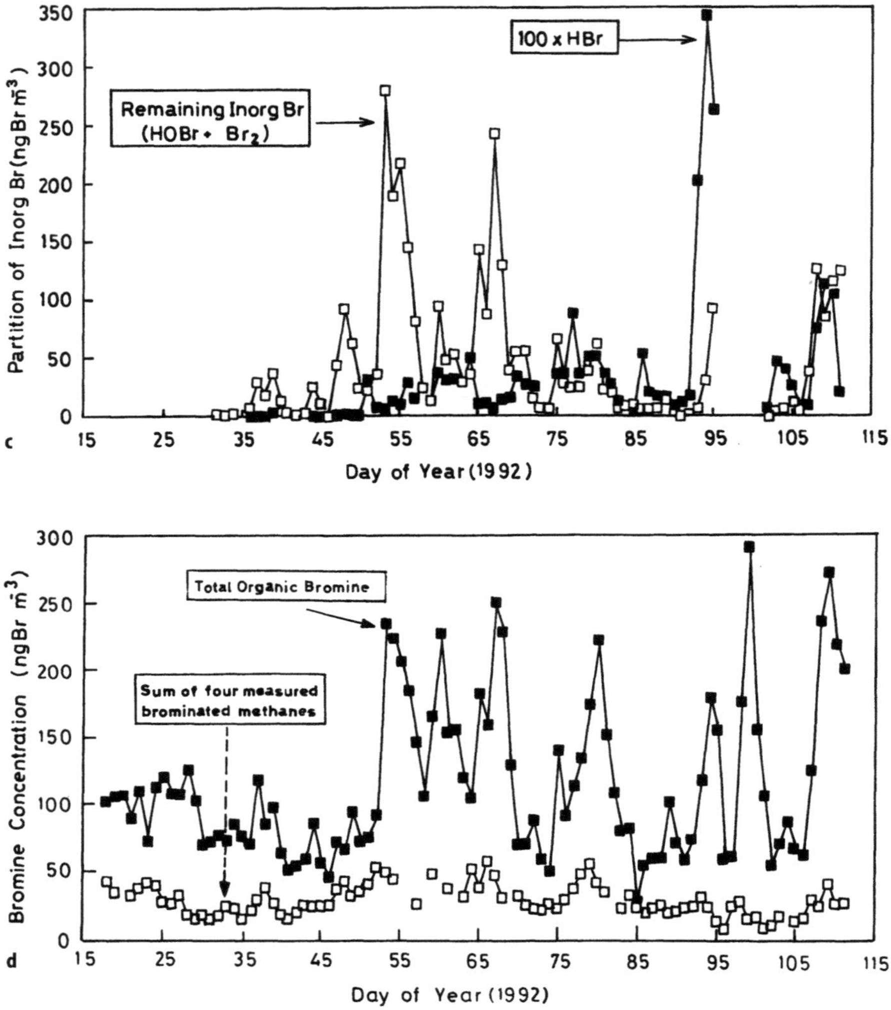

Fig. 8. **c** estimated HBr and the remaining InorgBr [89]; **d** total gaseous organic bromine (TorgBr) concentrations, indicated by the *solid squares* as determined from the FPC method. The sum of the four brominated methanes determined using the CHGC method is also plotted with the *open squares*[89]

this Arctic surface ozone hole formation [6, 90]. A strong anti-correlation between ozone and aerosal-bromine was observed for the whole period (Figs. 9 and 10a) similar to what has been observed earlier after polar sunrise (Figs. 9b, 10b [5, 76]. No correlation was observed with chlorine (Cl), but a week anti-correlation towards the end (in April) was found with iodine (I) [90]. After polar sunrise on day 65 (5 March, 1992), ozone depletion increased in magnitude. This is evident in Fig. 9a, with negative departures of ozone from free tropos-

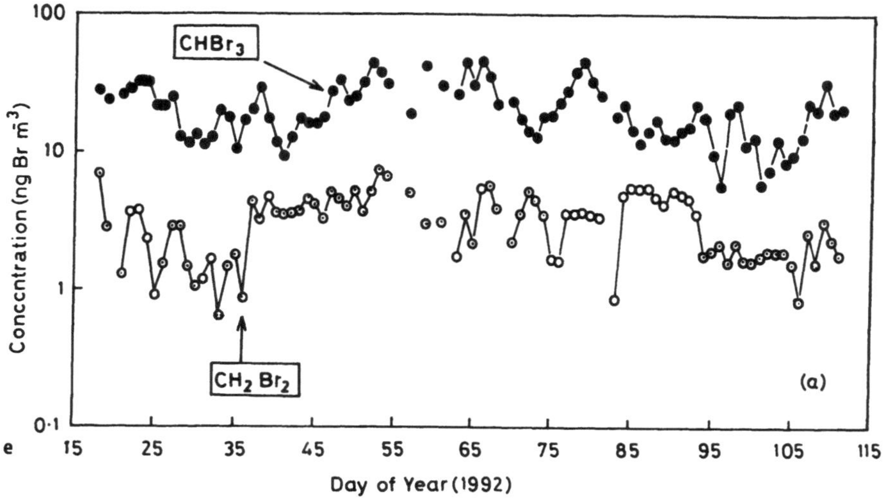

e

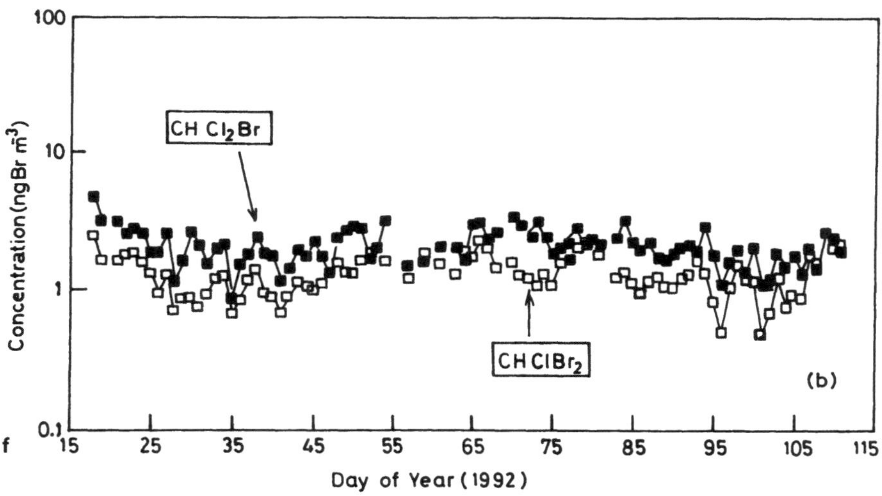

f

Fig. 8. e time series Br concentrations of $CHBr_3$ and CH_2Br_2 from the CHGC method [89]; f time series Br concentrations of $CHClBr_2$ and $CHCl_2Br$ from the CHGC method [89]

pheric levels of over 40 ppbv accompanied by distinct peaks of fine and coarse particulate Br [90].

Over 80% Br mass was as fine particles having less than 2.5 μm diameter, and unlike Cl, over 95% of both fine and coarse fractions, were non-sea salt Br [90]. It has been shown that Arctic aerosols contain sea-salt Br and automotive Br in equal though insignificant amounts [94]. Thus the Br measured during PSE-1992, must have originated from other sources, most likely from marine Br via

a

b

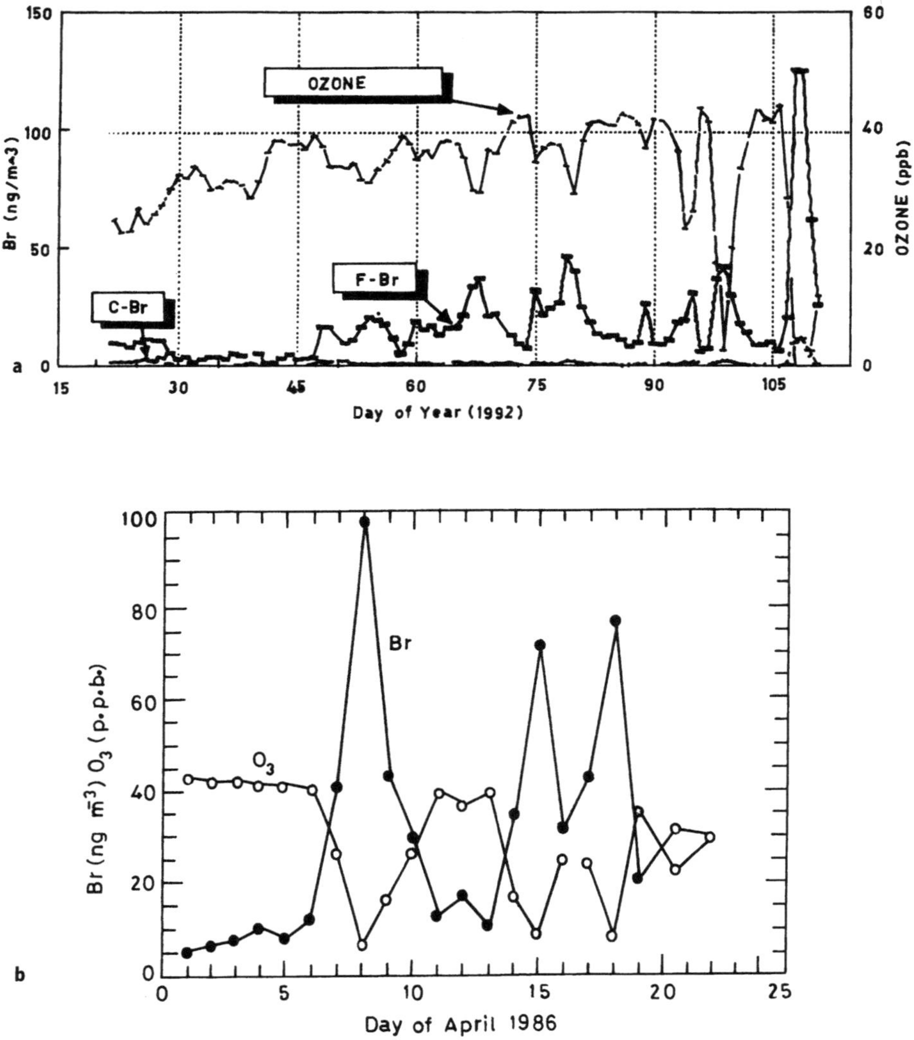

Fig. 9. **a** Daily fraction of fine and coarse bromine (F-Br and C-Br) compared with O_3 at Alert [90]; **b** comparison of daily concentrations of filterable Br and O_3 at Alert [5]

conversion into gaseous Br_2 or HBr and heterogeneous reactions on aerosols and ice with subsequent gas to particle conversion [93]. Simultaneous measurements of non-methane hydrocarbon composition [94] indicate that Br is more important than Cl in destroying ozone [6].

During PSE-1992 only bromine compounds showed a strong correlation with ozone destruction. All components of bromine - inorganic particulate, inorganic gaseous and organic gaseous [86, 89, 96] were anti-correlated with ozone, but the extent of ozone depletion differed. It increased from inorganic particulate to inorganic gaseous to organic gaseous Br. Inorganic Br appeared to be the sum of HBr, HOBr and Br_2 (Fig. 8b, c) whereas organic bromine consisted of the species CH_3Br, $CHBr_3$, CH_2Br_2, CH_2ClBr, $CHClBr_2$ and $CHCl_2Br$ (Fig. 8d–f). The measurements [89] carried out at Alert during January 18 to April 21, 1992 revealed the following:

(i) Particle bromide increased gradually from about 10 ng (Br) m^{-3} during the dark period to over 20 ng (Br) m^{-3} during the light period, with a marked

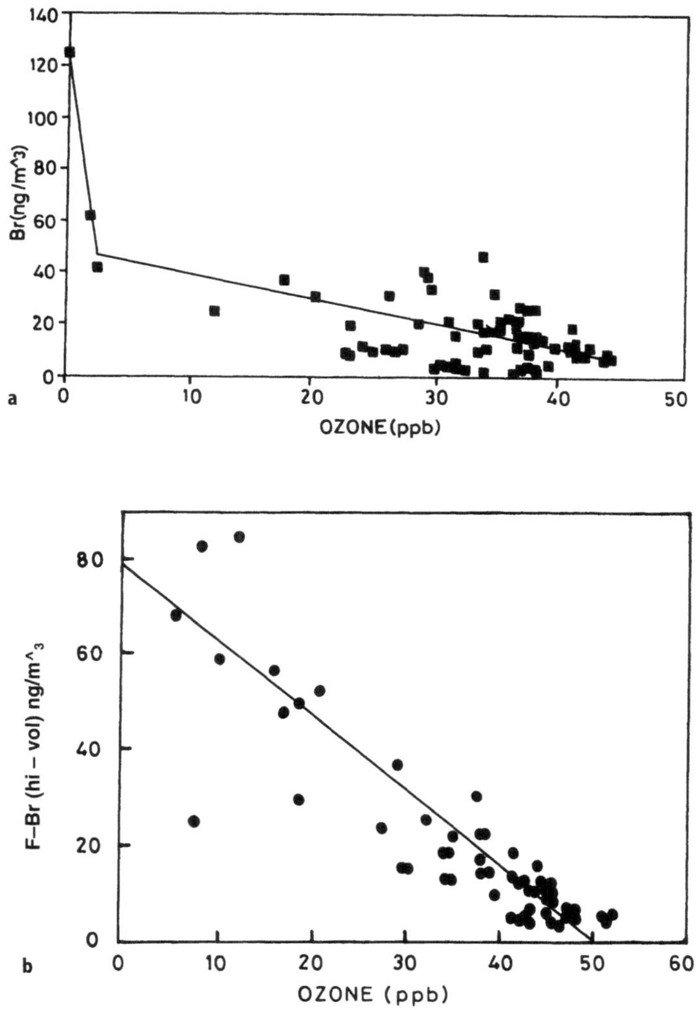

Fig. 10. **a** Correlation of fine fraction of Br and O_3 daily concentrations at Alert from Jan 20 to April 20, 1992 [90]; **b** regression of O_3 against filterable bromine(F-Br) at Alert in 1988 at Polar sunrise [76]

peak of 120 ng (Br) m^{-3} corresponding to a small ozone depletion event (Fig. 8a).

(ii) Inorganic gaseous bromine (InorgBr) was about 60 ng (Br) m^{-3} during the dark period and relatively constant. A major peak, up to 280 ng (Br) m^{-3} (Fig. 8b, c), before sunrise accompanied a similar peak in the total organic bromine. The episodes originated in the free troposphere over Greenland. After sunrise the peaks in inorganic Br corresponded to ozone depletion periods.

(iii) Total organic bromine was relatively constant before sunrise at 100 ng (Br) m^{-3} but more variable afterward, up to 280 ng (Br) m^{-3} (Fig. 8d). Individual species – $CHBr_3$ with levels of 7–60 ng (Br) m^{-3}, CH_2Br_2, CH_2ClBr, $CHClBr_2$ and $CHCl_2Br$ were lower at 0.5–7.5 ng (Br) m^{-3} (Fig. 8(e), (f)]. $CHBr_3$ was the largest contributor to the total organic bromine among these five species on an average accounting for 23%, while other four species amounted to less than 5% on average. CH_3Br, not measured during this study, was estimated to contribute 44% of the total organic bromine equivalent to a concentration of 40 ng (Br) m^{-3} or 11 pptv. The remaining contribution was probably from "missing" species which were episodically dominant after sunrise with concentrations up to 240 ng (Br) m^{-3} and possibly included some inorganic species. All the peaks in the organic bromines after sunrise corresponded to the ozone depletion event.

(iv) $CHBr_3$, $CHClBr_2$ and $CHCl_2Br$ were significantly correlated. The ratio $CHClBr_2/CHBr_3$ decreased linearly with increasing $\ln(CHBr_3)$, with a steeper decrease after sunrise than before. The decreases suggest different rate of destruction with $CHBr_3$ having larger rate constant than $CHClBr_2$. A similar relationship existed between the ratio $CHCl_2Br/CHClBr_2$ and in the $\ln(CHClBr_2)$, but the dark period slope was near zero, indicating a greater difference in rates in the two species in the light period.

During ozone depletion episodes, particulate Br was generally less than 10% of the total, except for the last event in April when it rose to 23% [92]. Mostly, as a result of more frequently occurring O_3 depletion episodes towards April, average concentrations of all Br components showed an increase. The sum of inorganic particulate and gaseous Br increased from about 10 to 20 pptv and organic gaseous Br from 25 to 40 pptv (Fig. 11) [92]. Formaldehyde measured along with $CHBr_3$, $CHBr_2Cl$, $CHBrCl_2$, showed elevated abundance too, with mixing ratios as high as 100–400 pptv in spring which enhanced further during ozone depletion episodes, suggesting that in-situ heterogeneous chemical reactions were involved in the regeneration of this NMHC-radical reaction product as well as of Br atoms, both of which would otherwise be quickly destroyed by reacting with each other [6]. Observed aerosol size distribution during these episodes, along with the findings that most of the accumulation mode mass of aerosol consisted of H_2SO_4 [92], indicated that ozone depletion in the lower Arctic troposphere (similar to that in the stratosphere), is likely to be associated with heterogeneous chemical reactions involving acidic sulphate aerosols [6].

Low level aircraft observations [88] indicated that these marked ozone epi sodes were not confined to the vicinity of Alert only but extended to larger areas

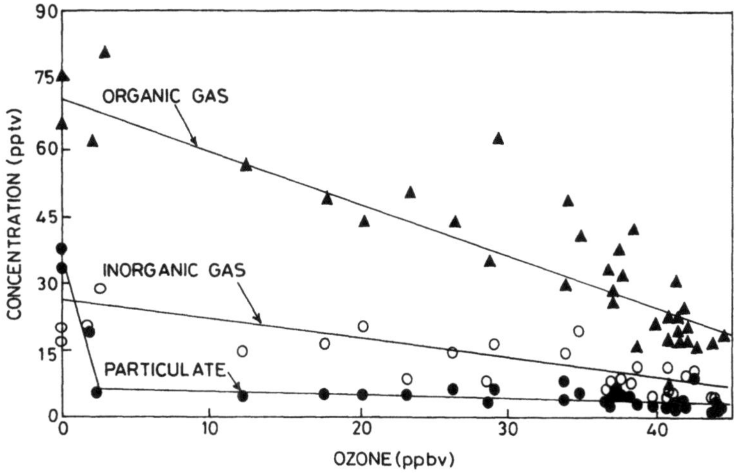

Fig. 11. The dependence of Br components on ozone for the period after polar sunrise (day of the year >75). The linear regression fits for two straight lines were [inorganic Br] = 22 – 0.38 [ozone], r = –0.81, N = 37; [organic Br] = 74 – 1.25 [ozone], r = –0.89, N = 36. The particulate Br line is drawn by eye [92]

of the Arctic. Further, the measurements [87] made at the ice camp SWAN (in the Lincoln Sea, 160 km Northwest of Alert) suggest that meteorology plays an important role by modulating different air masses: ozone depleted/halogen-rich air formed in the atmospheric boundary layer over the Arctic ocean near Alert alternating with free tropospheric ozone-rich/halogen poor air. Out on the Arctic ocean, ozone concentrations were depleted more and for longer periods than at the nearby Alert that receives a considerable amount of free tropospheric air via katabatic down slope flow [76, 6, 5].

4 Tropospheric Impact

Ozone, although existing in trace amounts only, plays a vital role in our atmosphere. It is of central importance in atmospheric chemistry because its photodissociation produces OH radicals, which, in turn, determine the oxidation efficiency of our atmosphere. About 90 percent of atmospheric ozone is found in the ozone layer between 10 and 40 km above the Earth's surface, and the rest in the troposphere (Fig 12 (1)). While ozone throughout the atmosphere is chemically identical, tropospheric ozone, however, displays a destructive side. Since ozone strongly reacts with other molecules, it can damage the living tissues of plants and animals. It is a key component of photochemical smog that hangs over many major cities across the world. Ozone is also a potential greenhouse gas, with an intense infra-red absorption band at 9.6 μm; changes in the amount and distribution of ozone can affect the radiative forcing on climate [101, 102].

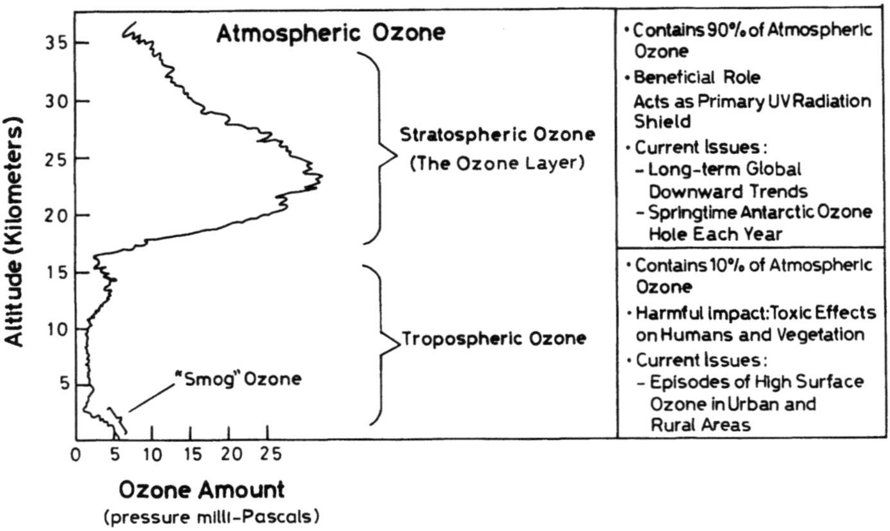

Fig. 12. Vertical distribution of ozone in the Earth's atmosphere [1]

However, unlike the stratospheric ozone depletion, ozone episodes cause no adverse effect.

The phenomenon of springtime tropospheric ozone destruction in the northern polar region is not of global significance but it is puzzling and to date not clearly understood.

A new combined BrO–ClO mechanism has been proposed to explain the Arctic ozone episodes [99] and is claimed to be more efficient than the BrO–BrO mechanism used by Barrie et al. [5]. Mixing ratios of BrO obtained in the range of 4–17 pptv [98], along with ClO concentrations estimated from Cl atom measurements in the Arctic troposhere, are being used to substantiate this mechanism. However, the verification will depend on precise ClO measurements to be achieved in the Arctic troposphere.

It may be pointed out that Cl atom concentrations derived from hydrocarbon data [5] imply only a small fraction of ozone destruction through Cl atom reaction. Although there is strong evidence for Cl atom chemistry in the removal of hydrocarbons, Cl atom chemistry alone cannot explain the ozone depletion episodes [5]. Simultaneous measurements of acetylene can be interpreted as the indication of additional removal by Br atoms. The extent of this effect implies that at temperatures higher than −30 °C, the observed ozone depletion at Alert can be attributed to ozone destruction by gas phase reactions with Br atoms [5].

An inverse correlation of ozone with bromine was clearly observed in March 1995 during PSE – 1995 [76a].

The photochemical mechanisms that are responsible for halogen activation have, so far, not been established. The catalytic activation of both Cl and Br radicals observed in the Arctic lower troposphere [98, 103, 103a] may consist of the following reaction sequences on the sea-ice surface [102a].

$$\begin{aligned} HOBr + Cl^- + H^+ &\rightarrow BrCl + H_2O \\ BrCl + h\nu &\rightarrow Br + Cl \end{aligned}$$

or

$$\begin{aligned} HOBr + Br^- &\rightarrow Br_2 + OH^- \\ Br_2 + h\nu &\rightarrow 2Br \end{aligned}$$

or

$$\begin{aligned} HOBr + Cl^- + H^+ &\rightarrow BrCl + H_2O \\ BrCl + Br^- &\rightarrow Br_2Cl^- \\ Br_2Cl^- &\rightarrow Br_2 + Cl^- \\ Br_2 + h\nu &\rightarrow 2Br \\ 2(Br + O_3 &\rightarrow (BrO + O_2) \\ 2(BrO + HO_2 &\rightarrow HOBr + O_2) \end{aligned}$$

$$\text{net: } 2HO_2 + H^+ + Br^- + h\nu \rightarrow HOBr + 4O_2 + H_2O$$

With HOBr and CL^- acting as catalysts.

According to Vogt et al. [103b], the last chain may be of particular importance.

5 Halons

Halons, $CBrClF_2$ (H-1211), $CBrF_3$ (H-1301) and $C_2Br_2F_4$ (H-2402), of totally anthropogenic origin with lifetimes of 20, 65 and 22 years respectively, are important bromine source gases.

CF_2Br_2 (H-1202), having volume mixing ratios around 0.03 pptv appears to be insignificant [18]. In 1991/1992 atmospheric abundance of H-1211, H-1301 and H-2402 were measured to be 2.58 pptv, 2.77 pptv and 1.41 pptv respectively at the tropopause [19]. In the same study it was estimated that 39% of the stratospheric organic bromine loading was more or less equally shared by these halons (other studies indicate that the abundance of H-2402 is between 0.2 and 0.5 pptv [18]). Ground based measurements of H-1211 and H-1301 carried out regularly during the same period (1991/1992) at seven monitoring stations: Alert, Barrow, Cape Grim, Mauna Loa, Niwot Ridge, Samoa (US) and South Pole, yielded volume mixing ratios of H-1211 little over 2 pptv and of H-1301 within 2 pptv [104]. Mixing ratios of H-1211 and H-1301 from other measurements are reported between 2 and 3 pptv and 1 and 2 pptv, respectively [1, 7, 18, 105–107]. Their background concentrations in the NH have increased, respectively, by a factor of 10 and 13 during 1978 and early 1995 [18].

Halons 1211 and 1301, are clean, stable, low toxic and very effective fire extinguishers. Halons are benign in the lower atmosphere and are transported to the stratosphere almost without decomposition. Their long atmospheric lifetimes also imply that halons already emitted will be releasing bromine in the stratosphere for several more decades after their production is stopped.

The concentrations of all the four halons (H-1211, H-1202, H-1301 and H-2402), as recorded in the Cape Grim archive, increased significantly since 1978, and all continue to rise [18], although their production has been stopped on 1st January 1994 by the Montreal Protocol on substances that deplete the ozone layer. However, for H-2402 there is some evidence for a slight reduction in growth rate since 1992 [18].

6 Stratospheric Bromine Budget and Chemistry

Polar stratospheric Ozone depletion, during winter/spring in both hemispheres, continues to be an important issue; while the Antarctic Ozone hole has become an annual event (with severe ozone depletion observed continuously for the last nine years, 1989–1997; Fig. (13a,b). During the past years increasing ozone losses have also been observed in the NH [102]. In the Arctic, the depletion have been smaller than those observed over Antarctica, and were consistent with the prediction of photochemical model calculations constrained by in-situ and satellite observations. Chlorine and bromine catalysed chemistry is identified as the cause for this ozone destruction [1, 7, 8, 12, 102, 105, 107].

Although stratospheric abundance of bromine source gases are less than 1% those of chlorine source gases, active bromine in the form of BrO plays an important role in photochemical ozone destruction in both hemispheres. All current assessments [1, 7, 8, 12, 102, 105–107] have indicated a greater potential of bromine than chlorine per atom to destroy ozone, particularly in the lower stratosphere. This is because about 50% of the available bromine is in reactive forms (Br and BrO) compared with chlorine which has only few percent in the reactive ClO form. Moreover, BrO primarily reacts synergistically with ClO in a fast catalytic cycle to destroy ozone even in the absence of oxygen atoms. Finally, bromine catalysis is most efficient in the lower stratosphere, where the ozone concentration is largest. As the abundance of chlorine increases throughout the stratosphere, ozone depletion by bromine catalysis increases in proportion with the increase in the abundance of both stratospheric chlorine and bromine [1]. Fig. 14 [109] shows the effectiveness of bromine in destroying ozone at mid

Fig. 13. **a** The historical signature of springtime total Ozone record for Antarctica as measured by Dobson spectrometers during October at Halley Bay, Syowa, and Faraday, and from 15–31 October at South Pole [1]; **b** area of the region enclosed by the 220 DU total ozone contour in the SH. The *white line* represents the 1978–1991 average with the *shaded area* representing the extremes for this period. The 1992 and 1993 areas are represented by the *continuous line* and *points*, respectively. 12 million square kilometers is about 5% of the surface area of the SH, so that the maximum extent of the region in 1992 or 1993 with total ozone less than 220 DU, if circular, was about 65 °S. Data for 1978–1992 are from Nimbus 7 TOMS; data for 1993 are from Meteor TOMS. Only measurements made south of 40 °S were considered, to avoid including any low tropical values recorded [1]. Deep and extensive ozone holes similar to these were also observed in the austral springs of 1994, 1995, 1996 and 1997 [102].

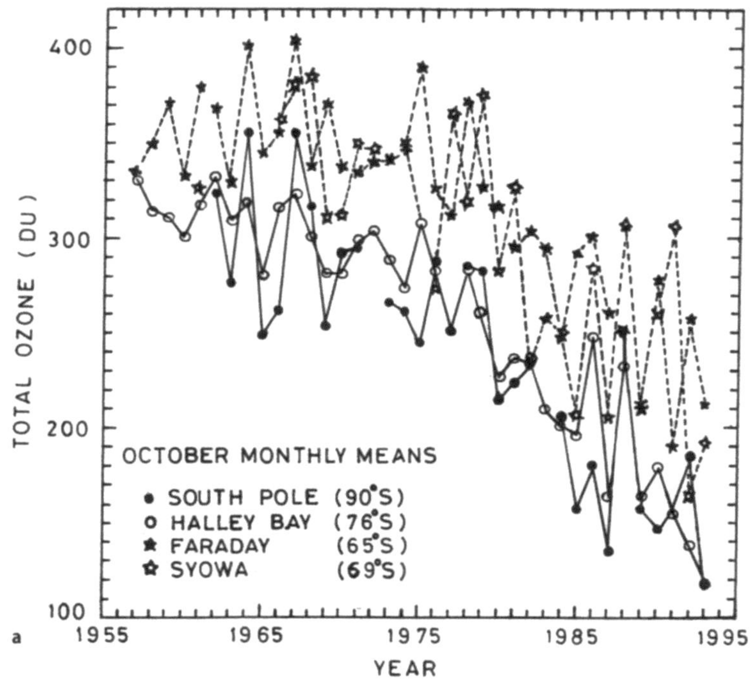

TOTAL OZONE (DU)
400
300
200
100
OCTOBER MONTHLY MEANS
SOUTH POLE (90°S)
HALLEY BAY (76°S)
FARADAY (65°S)
SYOWA (69°S)
1955
1965
1975
1985
1995
YEAR

a

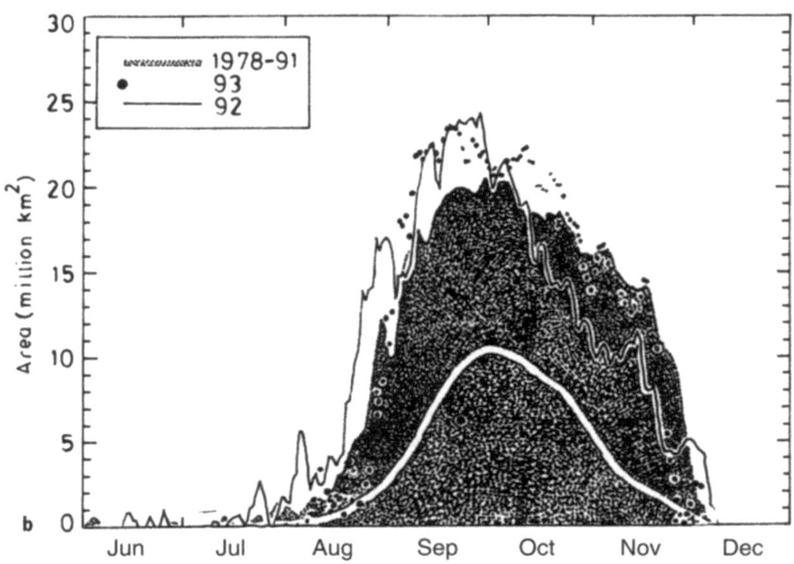

Area (million km²)
30
25
20
15
10
5
0
1978-91
93
92
Jun
Jul
Aug
Sep
Oct
Nov
Dec

b

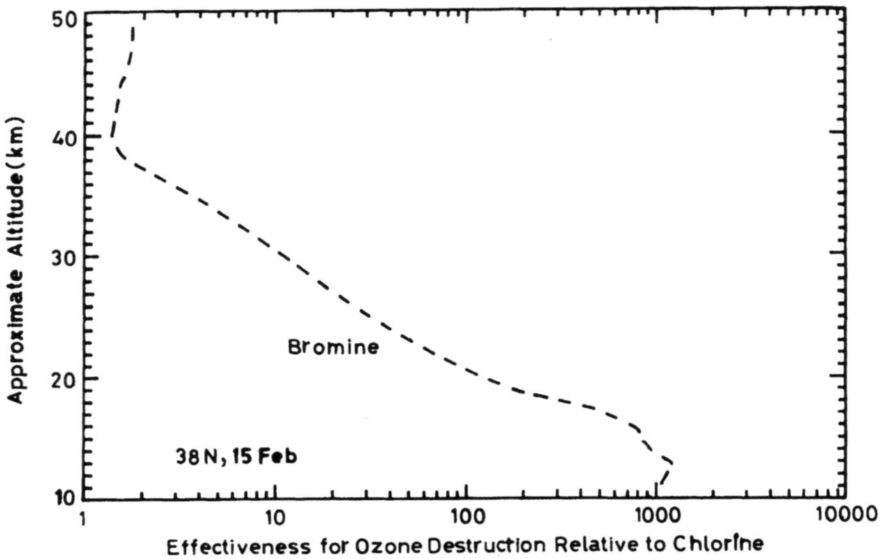

Fig. 14. Calculated effectiveness of bromine to destroy ozone at midlatitudes relative to chlorine [109]

latitudes relative to chlorine based on calculations with the zonally averaged chemical–radiactive–transport model of Lawrence Livermore National Laboratory (LLNL). Current chemistry in the model is based on recommendations of the NASA panel for data evaluation [110]. As can be seen from the figure, the bromine catalytic cycles are more than 100 times more efficient than the chlorine catalytic mechanisms in destroying ozone in the lower stratosphere, especially below 20 km. Similar results have also been obtained by Solomon et al. [111].

Heterogeneous chlorine reactions are important because they provide a mechanism, for rapid conversion of HCl into ClO_x (= Cl + ClO + $2Cl_2O_3$). Since the time scale for HCl production is several days, it is slowly recycled. Therefore, heterogeneous reactions considerably perturb the chlorine partitioning. In contrast, because the bromine species are short lived [112], heterogeneous bromine reactions are important as they allow the formation of catalytic cycles for the conversion of H_2O into HO_x (= OH + HO_2), HCl into ClO, and NO_x (= NO + NO_2) into HNO_3, as well as for the rapid recycling of the bromine reservoir species $BrONO_2$ and HBr. The hydrolysis of $BrONO_2$ on sulphate aerosols enhances the HOBr concentration, which in turn increases the OH and HO_2 abundance, thereby reducing the HCl lifetime and concentration. The enhancement in OH and HO_2 and reduction in HCl concentrations lead to additional ozone depletion at all latitudes and seasons. At temperatures below 210 K the bulk phase reaction of HOBr with HCl on sulphate aerosols becomes significant. The most important heterogeneous bromine reactions on PSCs are mixed halogen reactions of HCl with HOBr and $BrONO_2$, and HBr with HOCl and $ClONO_2$ [113].

Typical total organic bromine mixing ratios in the lower stratosphere are likely to be around 20 pptv, with nearly 54% contribution from CH_3Br, nearly 7–10% from CH_2Br_2 and the remaining (over 35%) shared by the three halons, H-1211, H-1301 and H-2402 [19]. Both remote and in-situ measurements of BrO indicate its mixing ratios between 4 and 10 pptv, approximately half the total organic bromine in the lower stratosphere [7, 32, 59, 61–66, 113 a]. Observations of high levels of OClO also confirm the BrO abundance since OClO is formed in the reaction ClO + BrO [64, 114–116]. Unfortunately, uncertainties in the measurements of BrO (± 30%) and in computations (± 80%) are large and, therefore, more measurements and analyses are needed before this issue can be resolved.

Upon reaching the stratosphere, bromine source gases (similar to chlorine source gases), decompose and liberate bromine atoms. Subsequent reactions, predominantly with O_3, OH, HO_2, ClO, NO and NO_2, partition inorganic bromine between reactive forms (Br and BrO), and reservoir forms ($BrONO_2$, BrCl, HOBr and HBr). The partitioning is fairly rapid, of the order of one hour or less, such that direct heterogeneous conversion of HBr and $BrONO_2$ to BrO is likely to have little impact on the partitioning of bromine [1]. The mixing ratio of HBr recently observed as 2.0 ± 0.8 pptv [67] in the range of 22 to 34 km and 1.31 ± 0.39 pptv between 20 and 36.5 km [113 b] altitudes suggest a small HBr branching ratio. Photochemical model studies and in-situ observations of the abundance and longitudinal gradients of BrO at mid-latitudes also indicate a similar ratio. Thus in all probability HBr is a minor reservoir for reactive bromine in the lower stratosphere and is unlikely to exceed the 2% of total inorganic bromine [1, 117]. Further, on the basis of laboratory measurements of rate constants and photolysis cross-sections, $BrONO_2$, HOBr and BrO are most likely the dominant reservoirs for inorganic bromine at mid-latitudes.

The stratospheric gas phase bromine cycle is shown in Fig 15 [1], and the three reaction cycles primarily responsible for the bromine catalysed ozone loss are given below [1].

$$\begin{aligned} ClO + BrO + h\nu &\rightarrow Br + Cl + O_2 \\ Br + O_3 &\rightarrow BrO + O_2 \qquad \ldots I \\ Cl + O_3 &\rightarrow ClO + O_2 \end{aligned}$$

$$\begin{aligned} BrO + HO_2 &\rightarrow HOBr + O_2\ I \\ HOBr + h\nu &\rightarrow OH + Br \qquad \ldots I \\ Br + O_3 &\rightarrow BrO + O_2 \\ OH + O_3 &\rightarrow HO_2 + O_2 \end{aligned}$$

$$\begin{aligned} BrO + NO_2 + M &\rightarrow BrONO_2 + MI \\ BrONO_2 + M &\rightarrow Br + NO_3 \\ NO_3 + h\nu &\rightarrow NO + O_2 \qquad \ldots II \\ Br + O_3 &\rightarrow BrO + O_2 \\ NO + O_3 &\rightarrow NO_2 + O_2 \end{aligned}$$

Cycles I and II are more important than Cycle III. In polar regions, where NO_x is reduced and ClO is enhanced by heterogeneous reactions on sulphate aerosols and polar stratospheric clouds, Cycle I dominates the ozone loss due to bromine. At mid-latitudes the first two cycles contribute approximately equally to ozone

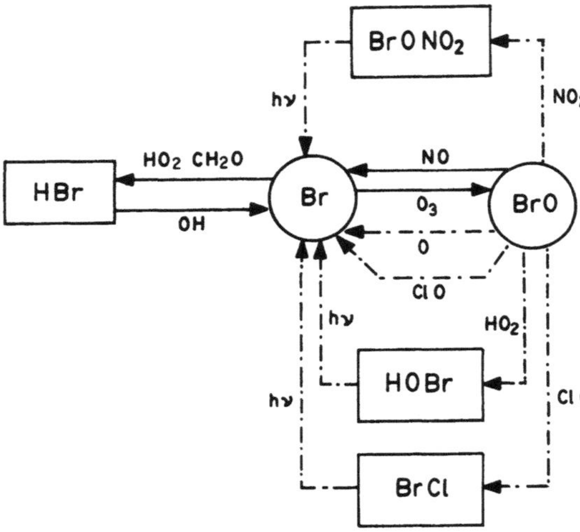

Fig. 15. Stratospheric gas phase bromine cycle [1]

loss at 20 km, and Cycle II is the most important near the tropopause, where the abundance of HO_2 is substantial but ClO abundance are negligible.

The importance of Cycle I in the lower stratosphere has been ascertained directly from simultaneous in-situ measurements of the abundance of BrO and ClO and concurrent ozone decreases within the Antarctic ozone hole [118]. Analyses using in-situ BrO and ClO data indicate that Cycle I contributed approximately 25% to ozone losses observed over Antarctica in 1987 [118, 119] and could contribute as much as 40% to total ozone loss over the Arctic in winter [114, 116].

On the other hand, due to uncertainties in the rate constant measurements of the BrO + HO_2 reaction at low temperatures, and in the measurements of HO_2 of about 50%, the importance of BrO+HO_2reaction is not so definite [62], compared to the ClO+BrO reaction where the uncertainties of these parameters are comparatively less [62, 118]. Moreover, in the perturbed polar regions where Cycle I dominates, uncertainties in HO_x kinetics and measurements are of little consequence to estimates of the importance of bromine to ozone losses [1].

The recent results from the BrO measurements during the Airborne Arctic Stratospheric Expedition (AASE-II) [66] are in contrast with the previously measured values of BrO in the stratosphere [62], where a significant increase was observed in the polar vortex relative to mid-latitudes. This raises the issue whether BrO at mid-latitudes during AASE-II was higher than that during previous campaigns, perhaps as a result of increased heterogeneous chemistry following the eruption of Mt. Pinatubo. Several studies have shown that, in the lower stratosphere, the addition of sulphate aerosol surface area had a noticeable effect on the abundance of NO_x and ClO [120–122]. Therefore, one might expect a decrease in NO_x to result in an increase in BrO analogous to what has been observed for ClO. However, when average BrO values as a function of latitude obtained before and after the eruption of Mt. Pinatubo were compared, three of the four available pairs showed little change and the fourth (at 20–30 °N, February 1992) showed a decrease in BrO. Thus without simultaneous measurements of organic and inorganic bromine species from these regions no conclusions can be drawn. However, if Br_yvalues were similar, this would imply that the mid-latitude BrO apparently did not respond to enhancement in sulphate aerosol surface area [66] in contrast to inferences from simultaneous ground based observations of BrO and NO_2 [65].

Heterogeneous reactions of bromine on aerosol and polar stratospheric cloud surfaces under simulated stratospheric conditions, have been studied in the laboratory [123–126]. When some of these reactions are incorporated in stratospheric photochemical models, a shift in the internal partitioning of bromine species in favour of HOBr, BrCl and Br_2 and a considerable reduction of $BrONO_2$ and HBr is observed. Further, with the assumption of a simultaneous activation of chlorine, a few weeks earlier during winter, via reduction of NO_x and additional consumption of HCl, an additional ozone depletion of about 30% is predicted under the volcanic conditions at the latitude of 60 °S and altitude corresponding to 50 mbar. Validation of these predictions through field measurements would shed some additional light on the present state of understanding.

7 Conclusions

In spite of a good amount of research work carried out during the last two decades, there still exist large uncertainties in the estimation of sources and sinks of atmospheric bromine bearing compounds. The current understanding of oceanic emission and uptake of CH_3Br leaves many open questions. Perhaps, never before so much confusion in this regard existed. Whether the ocean acts as a buffer towards changes in atmospheric CH_3Br, is an important question. If the ocean is a net sink, what are the sources of CH_3Br in the SH that result in an atmospheric methyl bromide concentration as high as 9 pptv?

The budget of atmospheric CH_3Br is reasonably constrained by current estimates of the atmospheric burden, lifetime and interhemispheric gradient; any description of the system must fit within the framework. The biggest issues facing scientists in the field are; first, balancing the sources and sinks while accounting for the observed interhemispheric gradient; second, determining how much anthropogenic emissions contribute to the total atmospheric budget; third, assessing the relative contribution of CH_3Br to the total stratospheric bromine; and, finally, evaluating the efficiency of bromine in depleting stratospheric ozone [17]. The last also requires some good direct measurements of total stratospheric bromine.

Intensive monitoring of short-lived bromine species, also beyond the Arctic region, identification and quantification of their sources, particularly anthropogenic, if any, better evaluation of their lifetimes and chemical mechanisms responsible for the Arctic ozone episodes, not at all clearly understood, are important issues. $CHBr_3$ and, to a lesser extent, CH_2Br_2 may enter the stratosphere episodically in regions of intense up-welling in the tropics; if so, these amounts need to be quantified.

The stratospheric bromine chemistry is now better understood, but the heterogeneous reactions on aerosol surfaces, especially over mid-latitudes in winter times need further clarification.

The ozone depletion potential of CH_3Br, an yardstick for policy regulation (in addition to its scientific importance), requires better estimation. The best values, 0.7 and 0.6 arrived at in the UNEP and WMO reports [1, 16,], are off by 30 to 45% or even more from the current estimate of less than 0.4.

However, before closing, the authors wish to emphasise that a lot has been accomplished. Recent research has indeed enhanced our understanding, and ongoing investigations addressing these issues are likely to resolve most of them.

Acknowledgements. ONS is grateful to Professor P. Fabian, for supporting his couple of stays at the Ludwig Maximilians University Munich, Freising Campus, Germany. He is also thankful to Professor P. Khastgir for his kind help.

8 References

1. WMO, Scientific Assessment of Ozone Depletion: 1994, World Meteorological Organization Global Ozone Research Monitoring Project Report No. 37, 1995
2. Yvon SA, Butler JH (1996) Geophys Res Lett 23:53
3. Farman JC, Gardiner BG, Shanklin JD (1985) Nature 315:207
4. Chubachi S (1984) Mem Natl Inst Polar Res 34:13
5. Barrie LA, Bottenheim JW, Schnell RC, Crutzen PJ, Rasmussen RA (1988) Nature 334:138
6. Barrie LA, Bottenheim JW, Hart WR (1994) J Geophys Res 99:25, 313
7. Singh ON (1992) In: W A Nierenberg (ed) Encyclopedia of earth system science, Vol 1, Academic Press, NY, p 135
8. Singh ON, Fabian P (1995) Speculations Sci Technol 18:9
9. Daniel JS, Solomon S, Albritton DL (1995) J Geophys Res 100:127
10. Lobert JM, Butler JH, Montzka SA, Geller LS, Myers RC, Elkins JW (1995) Sci 267:1002
11. Butler JH (1994) Geophys Res Lett 21:185
12. WMO, Scientific Assessment of Ozone Depletion: 1991, World Meterological Organization Global Ozone Research and Monitoring Project -Report No. 25, 1992
13. Singh HB, Kanakidou M (1993) Geophys Res Lett 20:133
14. Khalil MAK, Rasmussen RA, Gunawardena R (1993) J Geophys Res 98:2287
15. Reeves CE, Penkett SA (1993) Geophys Res Lett 20:1563
16. Albritton DL, Watson RT (1992) In: Watson RT, Albritton DL, Anderson SO, Lee-Bapty S (eds) Methyl bromide: its atmospheric science, technology and economics, Montreal Protocol Assessment Supplement, United Nations Environment Programme, Nairobi, Kenya
17. Butler JH (1995) Nature 376:469
18. WMO (1997) Consultation of Experts on Reactive Halogen Compounds and their Possible Effect on Ozone: 1995, World Meteorological Organization Global Ozone Research Monitoring Project Report No. 40
19. Schauffler SM, Heidt LE, Pollock WH, Gilpin TM, Vedder JF, Solomon S, Lueb RA, Atlas EL (1993) Geophys Res Lett 20:2567
20. Yagi K, Williams J, Wang NY, Cicerone RJ (1993) Proc Natl Acad Sci US 90:8420
21. Mano S, Andreae MO (1994) Sci 263:1255
22. Cicerone RJ, Heidt LE, Pollock WH, (1988) J Geophys Res 93:3745

23. Methyl Bromide State of the Science Workshop, Monterey, CA, USA, 1995
24. Singh HB, Salas LJ, Stiles RE (1983) J Geophys Res 88:3684
25. Penkett SA, Jones BMR, Rycroft MJ, Simmons DA (1985) Nature 318:550
26. Blake DR, Sive B, Zondlo M, Rowland FS (1993) EOS Trans AGU 74 (43):13.
27. Lobert JM, Butler JH, Montzka SA, Geller LS, Myers RC, Elkins JW (1995) EOS Trans AGU 76:S168
28. Lobert JM et al. (1997) Geophys Res Lett 24:171
29. Montzka SA, Butler J, Elkins J, Yvon S, Clarke A, Lobert J, Lock L (1995) EOS Transact AGU 76:160
30. Pilinis C, King DB, Saltzman ES, Penner JF, Watson JJ (1995) In 'Methyl Bromide State of the Science Workshop' Monterey, CA, USA
31. Lal S, Borchers R, Fabian P, Patra PK, Subbaraya BH (1994) Tellus 46B:373
32. Fabian P, Borchers R, Kourtidis K (1994) Geophys Res Lett 21:1219
33. UNEP (1995) 1994 Report of the Methyl Bromide Technical Options Committee Assessment
34. Thomas W (1996) Atmos Environ 30:p i-iv
35. Pilinis C, King DB, Saltzman ES (1996) Geophys Res Lett 23:817
35a. Yvon-Lewis SA, Butler JH (1997) Geophys Res Lett 24:1227
36. Crutzen PJ, Heidt LE, Krasnec JP, Pollock WH, Seiler W (1979) Nature 282:253
37. Bowen HJM (1979) Environmental Chemistry of Elements, Academic Press, London
38. Andreae MO, et al. (1996) J Geophys Res 101:23603
39. McKenzie LM, Ward DE, Hao WM, In: Levine JS (ed) Biomass burning and global change. MIT Press, Cambridge, Mass (submitted in 1995)
40. Penner JC, Liousse C, Pilinis C (1995) In: Methyl bromide state of the science workshop. Monterey, CA USA
41. Blake DR, Rowland FS (1995) In: Methyl bromide state of the science workshop. Monterey, CA USA
41a. Thomas VM, Bedford JA, Cicerone RJ (1997) Geophys Res Lett 24:1371
42. Mellouki A, Talukdar RK, Schmolter A-M, Gierczak T, Mills MJ, Solomon S, Ravishankara AR, (1992) Geophys Res Lett 19:2059
43. Zhang Z, Saini RD, Kurylo MJ, Huie RE (1992) Geophys, Res, Lett 19:2413
44. Prinn RG, et al. (1992) J Geophys Res 97:2445
45. Prather MJ, Spivakovsky CM (1990) J Geophys Res 95:18723
46. Talukdar RK, Mellouki A, Schmoltner A-M, Watson TJ, Montzka S, Ravishankara AR (1992) Science 257:227
47. Prinn RG, Weiss RF, Miller BR, Huang J, Alyea FM, Cunnold DM, Fraser PB, Hartley DE, Simmons PG (1995) Science 269:187
48. Shorter JH, Kolb CE, Crill PM, Kerwin RA, Talbot RW, Hines ME, Harriss RC (1995) Nature 377:717
49. Oremland R, Visscher P, Guidetti J, Connel T, Miller L (1995) In "Methyl Bromide State fo Science Workshop", Monterey, CA, USA
50. Montzka S, Butler J, Lobert J, Myers R, Clake A, Elkins J (1995) In 'Methyl Bromide State of the Science Workshop', Monterey, CA, USA
51. King DB, Saltzman ES (1999) J Geophys Res (in press); Saltzman S, DeBruyn WJ (1997) Marine Chem 56:51
52. Elliott S, Rowland FS (1993) Geophys Res Lett 20:1043
53. Elliott S, Rowland FS (1995) J Atmos Chem 20:229
54. Jeffers PM, Wolfe NL (1996) Geophys Res Lett 23:1773
55. Oremland RS, Miller LG, Strohmaier FE (1994) Eviron Sci Technol 28:514
56. Ravishankara AR, Lovejoy ER (1994) J Chem Soc Faraday Trans 90:2159
57. Solomon S, Mills M , Heidt LE, Pollock WH, Tuck AF (1992)J Geophys Res 97:825
58. Wuebbles DJ (1983) J Geophys Res 88:1433
59. WMO, Scientific Assessment of Stratospheric Ozone: 1989, World Meteorological Organization Global Research and Monitoring Project Report No. 20, Geneva 1990
60. Solomon S, Albrittion DL (1992) Nature 357:33
61. Carroll MA, Sanders RW, Solomon S, Schmeltekopf AL (1990) J Geophys Res 95:16633

62. Toohey DW, Anderson JG, Brune WH, Chan KR (1990) Geophys Res Lett 17:513
63. Wehner A, Callies J, Dorn H-P, Platt U, Schiller C (1990) Geophys Res Lett 17:517
64. Wahner A, Schiller C (1992) J Geophys Res 97:8047
65. Arpag KA, Johnson PV, Miller HL, Sanders RW, Solomon S (1994) J Geophys Res 99: 8175
66. Avallone LM, Toohey DW, Schauffler SM, Pollock WH, Heidt LE, Atlas EL, Chan KR (1995) Geophys Res Lett 22:831
67. Johnson DG, Traub WA, Chance KV, Jucks KW (1995) Geophys Res Lett 22:1373
68. Ko MKW et al. (1997) J Geophys Res 102:25507
69. Georgiev AA, Selwyn MR, Georges JE (1995) In "Methyl bromide state of science workshop" Monterey, CA, USA
70. Kourtidis K, Borchers R, Fabian P (1996) Geophys. Res. Lett. 23:2581
71. Rasmussen RA, Khalil MAK (1984) Geophys Res Lett 11:433
72. Penkett SA, Jones BMR, Rycroft MJ, Simmons DA (1985) Nature 318:550
73. Khalil MAK, Rasmussen RA (1985) Antarctic J US 19:206
74. Berg WW, Heidt LE, Pollock W, Sperry PD, Cicerone RJ (1984) Geophys Res Lett 11:429
75. Class TR, Ballschmiter K (1988) J Atmos Chem 6:35
76. Bottenheim JW, Barrie LA, Atlas E, Heidt LE, Niki H, Rasmussen RA, Shepson PB (1990) J Geophys Res 95:18, 555
76a. Impey GA, Shepson PB, Hastie DR, Barrie LA (1997) J Geophys Res 102:16005
77. Schall C, Heumann KG (1993) Fresenius J Anal Chem 346:717
78. Sturges WT, Cota GF (1995) In "Naturally-produced organohalogens" (ed) A Grimvall and EWB deLeer, Kluwer Academic Publishers, The Netherlands, p 385
79a. Yokouchi Y et al. (1997) J Geophys Res 102:8805
79. Yokouchi Y, Barrie LA, Toom D, Akimoto H (1996) Atmos Environ 30:1723
80. Gschwend PM, MacFarlane JK, Newman KA (1985) Science 227:1033
81. Sturges WT, Cota GF, Buckley PT (1992) Nature 358:660
82. Tokarczyk R, Moore RB (1994) Geophys Res Lett 21:285
83. Sturges WT, Sullivan CW, Schnell RC, Heidt LE, Pollock WH (1993) Tellus 45B:120
84. Sturges WT, Cota GF (1997) Marine chemistry 56:181
85. Moore RM, Tokarczky R, Tait VK, Poulin M, Geen C (1995) In: Naturally-produced Organohalogens (ed) A Grimvall, EWB de Leer, Dordrecht; Kluwer Academic Publishers, The Netherlands
86. Yokouchi Y, Akimoto H, Barrie LA, Bottenheim JW, Anlauf K, Jobson BT (1994) J Geophys Res 99:25, 379
87. Hopper JF, Peters B, Yokouchi Y, Niki H, Jobson BT, Shepson PB, Muthuramu K (1994) J Geophys Res 99:25, 489
88. Leaitch WR, Barrie LA, Bottenheim JW, Li SM, Shepson PB, Muthuramu K, Yokouchi Y (1994) J Geophys Res 99:25, 499
89. Li S-M, Yokouchi Y, Barrie LA, Muthuramu K, Shepson PB, Bottenheim J W, Sturges W T, Landsberger S (1994) J Geophys Res 99:25, 415
90. Barrie LA, Staebler R, Toom D, Georgi B, den Hartog G, Landsberger S, Wu D (1994) J Geophys Res 99:25, 439
91. Barrie LA, DenHartog G, Bottenheim JW, Landsberger S (1989) J Atom Chem 9:101
92. Barri LA, Li S-M, Toom DL, Landsberger S, Sturges W (1994) J Geophys Res 99:25, 453
93. McConnel JC, Henderson GS, Barrie L, Bottenheim J, Niki H, Longford CH, Templeton EMJ (1992) Nature 355:150.
94. Sturges WT, Barrie LA (1988) Atmos Environ 22:1179
95. Finlayson-Pitts BJ, Livingston FE, Berko NH (1990) Nature 343:622
96. Oltmanss J et al. (1989) In: Bojkov RD, Fabian P (eds) Ozone in the atmosphere. A Deepak Publ, Hompton, USA, p 498
97. Muthuramu K, Shepson PB, Bottenheim JW, Jobson BT, Niki H, Anlauf KG (1994) J Geophys Res 99:25, 369
98. Hausman M, Platt U (1994) J Geophys Res 99:25,399
99. Le Bras G, Platt U (1995) Geophys Res Lett 22:599

100. Anlauf KG, Mickle RE, Trivett NBA (1994) J Geophys Res 99:25, 345
101. Singh ON (1991) J Sci Res BHU (Platinum Jubilee Issue) 41 c:17
102. Singh ON (1998) Proc Ind Nat Acad Sci A-64:149
102a. Andreae MO, Crutzen PJ (1997) Sci 276:1052
103. Jobson BT, Niki H, Yokouchi Y, Bottenheim J, Hopper F, Leaitch R (1994) J Geophys Res 99:25, 355
103a. Finlayson-Pitts BJ, Pitts Jr JN (1997) Sci 276:1045
103b. Vogt R, Crutzen PJ, Sander R (1996) Nature 383:327
104. Elkins J W et al. (1994) In Boden TA, Kaiser DP, Sepanski RJ, Stoss FW (eds) Trends 1993: A compendium of data on global change, p 432–471
105. Singh O N, Borchers R, Fabian P, Lal S, Subbaraya B H (1988) Nature 334:593
106. Climate Change 1995, (ed) J T Houghton et al., Cambridge University Press, 1996
107. Singh ON, Borchers R, Lal S, Subbaraya BH, Kruger BC, Fabian P (1994) In Hudson R (ed) Ozone in Troposphere and Stratosphere. NASA Publ CP # 3266, p 827
108. Montzka SA et al (1996) Science 272:1318
109. Wuebbles DJ, Connell PS, Patten KO (1995) In Miziolek AW, Tsang W (eds) Halon Replacements: Technology and Science. The American Chemical Society, p 59
110. Demore WB et al. (1992) Kinetics and Photochemical Data for Use in Stratospheric Modelling, Evaluation #10, Jet Propulsion Lab, Pasadena, CA
111. Solomon S, Burkholder JB, Ravishankara AR, Gracia RR (1994) J Geophys Res 99:20, 929
112. Lary DJ (1996) J Geophys Res 101:1505
113. Lary DJ, Chipperfeld MP, Toumi R, Lenton T (1996) J Geophys Res 101:1489
113a. McKinney KA, Pierson JM, Toohey DW (1997) Geophys Res Lett 24:853
113b. Nolt IG et al. (1997) Geophys Res Lett 24:281
114. Salawitch RJ, Wofsy SC, McElroy MB (1988) Planet Space Sci 36:213
115. Salawitch RJ et al. (1993) Science 261:1146
116. Sessler J, Chipperfield MP, Pyle JA, Toumi R (1995) Geophys Res Lett 22:687
117. Gracia RR, Solomon S (1994) J Geophys Res 99:12, 937
118. Anderson JG, Brune WH, Lloyd SA, Toohey DW, Sander SP, Starr WL, Loewenstein M, Podolske JR (1990) J Geophys Res 95:11, 480
119. Murphy DM (1991) J Geophys Res 96:5045
120. Fahey DW et al. (1993) Nature 363:509
121. Wilson J C et al. (1993) Sci 261:1140
122. Avollone LM, Toohey DW, Proffitt MH, Margiton JJ, Chan KR, Anderson JG (1993) Geophys Res Lett 20:2519
123. Abbatt JPD (1994) Geophys Res Lett 21:665
124. Hanson DR, Ravishankara AR (1995) Geophys Res Lett 22:385
125. Williams LR, Golden DM, Huestis DL (1995) J Geophys Res 100:7329
126. Danilin MY, McConnel JC (1995) J Geophys Res 100:11, 237

CHAPTER 2

Reactive Chlorine Compounds in the Atmosphere

M. A. K. Khalil

Department of Physics, Portland State University, P.O. Box 751, Portland, Oregon 97207–0751, USA. *E-mail: aslam@atmos.phy.pdx.edu*

The defining characteristic of reactive chlorine gases is that they are removed by tropospheric processes such as oxidation, hydrolysis in clouds and oceans, rain, or by deposition. Chlorine-containing gases may be classified according to their lifetimes in the troposphere, which is often a reflection of their chemical reactivities. If the lifetime is less than about one year, the gases are considered to be "reactive"; if it is between 1–10 years, the gases are "moderately reactive"; and if the lifetimes are longer, the gases are "unreactive" and are not removed in the lower atmosphere. There are some 23 Tg of gaseous chlorine in the atmosphere or about 4 parts per billion of the troposphere. Of this, only about 3 Tg can be identified to come from natural sources, and all of it is in the form of reactive chlorine gases. Moderately reactive and unreactive gases are entirely man-made. This chapter is a review of the global budgets and trends of chlorine-containing gases with emphasis on the reactive group. All the known gases in each group are included here if concentrations are above a few parts per trillion and the existence of the gas is mostly due to direct emissions. The group of reactive gases consists of chloromethane or methyl chloride (CH_3Cl), chloroform ($CHCl_3$), phosgene ($COCl_2$), dichloromethane (CH_2Cl_2), chlorinated ethylenes (C_2HCl_3, C_2Cl_4), chlorinated ethanes (CH_4Cl_2, $C_2H_2Cl_4$), some other rarer gases, and inorganic chlorine gases such as hydrogen chloride (HCl) and molecular and atomic chlorine (Cl_2, Cl). Natural sources are identified for most of these gases. Methylchloride, for instance, comes from the oceans and fungal action on rotting wood. Although methylchloride is the most abundant chlorine-containing gas in the atmosphere, its budget is not understood at present. The class of moderately reactive gases consists of methylchloroform (CH_3CCl_3), chlorodifluoromethane ($CHClF_2$), dichlorofluoromethane ($CHCl_2F$), and other less abundant chlorinated gases with concentrations below a few parts per trillion. The long-lived or unreactive gases are the chlorofluorocarbons (CCl_2F_2, CCl_3F, $C_2Cl_3F_3$, $C_2Cl_2F_4$, C_2ClF_5, CCl_3F) and carbon tetrachloride (CCl_4). The long-lived gases are discussed here for comparison and because of their important role in the global cycle of gaseous chlorine. The most important gases in these last two groups have well documented anthropogenic emissions that are consistent with the observed concentrations, and no natural sources have been unequivocally identified. Since these gases will no longer be produced because of the Montreal Protocol and its amendments (international agreements to phase out global production of compounds that can deplete the stratospheric ozone layer), new compounds are being manufactured to take their place. The first data on some of the new chlorinated replacement compounds, namely HCFC-141b and HCFC-142b, are discussed.

Keywords: Reactive chlorine, Methylchloride, Chloroform, Phosgene, Dichloromethane, Trichloroethylene, Perchloroethylene, Chlorofluorocarbons.

The Handbook of Environmental Chemistry Vol. 4 Part E
Reactive Halogen Compounds in the Atmosphere
(ed. by P. Fabian and O. N. Singh)

1 Introduction

Much of what we know about atmospheric chlorine-containing gases has been driven by the idea, proposed more than two decades ago, that chlorine atoms can catalytically destroy stratospheric ozone. The identification of chlorofluorocarbons as one of the principal agents for transporting chlorine to the stratosphere focused attention on emissions from industrial processes and consumer goods as sources of global pollution (Molina and Rowland 1974; WMO 1985, 1988, 1989, 1991, 1995). The increasing concentrations of the chlorofluorocarbons reduce stratospheric ozone, potentially leading to higher levels of biologically harmful ultraviolet radiation at the earth's surface. It has been known all along that there are many naturally occurring chlorine-containing gases, and there are anthropogenic sources, other than the chlorofluorocarbons, that also contribute chlorine to the atmosphere. With very few exceptions, the budgets of these more reactive gases are still not understood.

Reactive chlorine gases are removed mostly in the troposphere by various processes such as atmospheric oxidation reactions, hydrolysis in cloud water or the oceans, heterogeneous reactions with atmospheric aerosols, and deposition directly on the earth's surface or in rain. The atmospheric lifetimes of these gases tend to be relatively short, ranging from a few days to several years. This is because the troposphere is large, well mixed, and chemically active. For most gases of interest here, reactions with hydroxyl radicals are the main removal process, and the lifetimes of the reactive chlorine gases are less than 10 years. Gases not removed or destroyed by tropospheric processes tend to have much longer lifetimes.

Reactive chlorine has several important roles in the environment as pointed out by Graedel and Keene (1995). It can contribute to the acidity of precipitation in remote areas (Galloway et al. 1982), it may remove hydrocarbons and other compounds when cycled as atomic chlorine radicals (Keene et al. 1996), and it may cause corrosion of metals (Rice et al. 1980; Graedel and McGill 1986). Because there are substantial natural emissions of reactive chlorine into the atmosphere that exceed the emissions from human activities, there have been doubts about the importance of man-made chlorofluorocarbons in the global environment. However, most of the reactive chlorine that is emitted does not reach the stratosphere, and that which does has competing destruction pathways that can reduce its impact on the ozone layer.

Chlorine-containing gases can be classified according to estimated reactivity into three groups: gases with atmospheric lifetimes of a few days to one year (reactive), with lifetimes of 1–10 years (moderately reactive), and gases with lifetimes longer than 10 years (long-lived).

2 The Budgets of Reactive Chlorine-Containing Gases

2.1 Budgets of Gases

The mass balance of a trace gas at time t, in an infinitesimal region dV of the atmosphere, is written as:

$$\frac{\partial C}{\partial t} = S - \frac{C}{\tau} - \frac{1}{n}\left[\underline{\nabla} \cdot n(\underline{v} - K \cdot \underline{\nabla})\right]C \qquad (1)$$

where C is the mixing ratio (pptv), S represents the emissions of the gas from its various sources (pptv/s), τ is the lifetime (s), which may be a composite of many different first-order processes, K is the transport coefficient (cm^2/s), v is the velocity of the mean winds (cm/s), and n is the number density of air (molecules/cm^3). In practice it is often convenient, and sufficient, to use spatially averaged versions of this equation (see Khalil and Rasmussen 1984). Averaging this equation over the entire atmosphere gives:

$$\frac{dC(t)}{dt} = S(t) - \frac{C}{\tau(t)} - \frac{C(t)}{\tau(t)} \qquad (2)$$

This equation is useful in providing a perspective and a check for mass balances of gases. According to these relationships, the mass balance consists of three elements: concentration, source, and sink. Most results quoted here are based on such models with various scales of spatial resolution and explicit descriptions of key physical and chemical processes.

If the lifetime is very short, the gas is not transported far from its source, thus, according to Eq. 1, the observed concentration is determined by local sources and sinks, as these can greatly exceed the transport term. By extension, for reactive gases, even those with modest atmospheric lifetimes, concentrations will tend to follow the regions of sources and sinks. High concentrations will be observed near source regions or away from the sinks. Gases with large sources from industrialized countries have higher concentrations in the middle and high latitudes of the northern hemisphere and lowest concentrations in the southern hemisphere. This is particularly true for many of the reactive chlorine gases as the emissions are high in the latitudes between 30–50° north, and, at the same time, the destruction, if it is due to reactions with OH radicals, is generally slow compared to the tropics (see e.g. Lu and Khalil 1991). Vertical mixing is sufficiently rapid that the concentrations of reactive gases do not decrease appreciably up to the tropopause if the lifetime is longer than a few months.

2.2
Nature of the Data Available

As discussed earlier, the data needed for the budgets consist of concentrations, emissions, and sinks. The most abundant data are measurements of concentrations in air. The surface data are classified into three groups: 1) systematic global time series of measurements (applicable to CH_3Cl, $CHCl_3$, CH_3CCl_3, $CHClF_2$, CCl_3F, CCl_2F_2, CCl_4, and $C_2Cl_3F_3$); 2) data from ship cruises lasting a few months (applicable to CH_2Cl_2, C_2HCl_3, C_2Cl_4, $C_2H_4Cl_2$, $C_2H_2Cl_4$, $C_2Cl_2F_4$, C_2ClF_5, and most compounds in the previous group); and 3) sporadic measurements by one or more research groups (applicable to HCl, $COCl_2$, CH_2Cl_2, $CHCl_2F$, C_2HCl_3).

Although "the next best thing", ship cruise or aircraft data can contain much less information compared to the systematic time series. On the other hand, such data can cover more latitudes than systematic measurements, particularly for the southern hemisphere where few locations are routinely sampled. The main problem with cruise data is the distortion of the latitudinal distribution because of seasonal variations of trace gases and local emissions, particularly for gases with lifetimes of less than a year. Such distortions are lessened if the measurements are taken in a transition season.

In addition to surface data, there are measurements throughout the vertical extent of the troposphere and the stratosphere. In the boundary layer, measurements are taken using tethered balloons, aircraft, or on top of mountains that are often above the boundary layer. For the free troposphere, measurements are usually taken on research aircraft and constitute instantaneous concentrations of gases. For short-lived species with sizable sources, these measurements do not reflect the mean state of atmospheric composition, especially at low latitudes. Stratospheric samples are taken with high-altitude balloons, rockets, and sometimes with high-flying aircraft. The stratospheric data we will use here have been taken using balloons, which extend the measurements up to about 35 km, covering most of the stratosphere (Fabian et al. 1981 a, b; 1984; 1985 a, b; 1996; Fabian and Gomer 1984; Goldan et al. 1980; Lal et al. 1985). Above these altitudes the gases of interest are very rare.

There are considerably fewer data on emissions. These are classified into two groups: 1) industrial sources for which the most accurate data are available from the production and sales records (applicable to CFCs, CH_3CCl_3, CCl_4) and 2) direct measurements of emissions from the oceans, soils, agricultural systems, and plants (applicable to CH_3Cl, $CHCl_3$, C_2Cl_4, C_2HCl_3). To achieve a proper mass balance, the observed concentrations, measured emissions and sinks must satisfy Eqs. 1 and 2.

For sinks, there are direct measurements mostly for the oceans and in some cases for soils (applicable to CH_3Cl, $CHCl_3$, and many CFCs). The role of soils and the biota, in either emitting or consuming reactive chlorine gases, is essentially unknown. For atmospheric sinks, reaction rates and photochemical data are available for most gases considered here (NASA 1997).

The discussion of the individual gases in subsequent sections is organized according to the classification of "reactive", "moderately reactive", and "long-lived" as discussed earlier. Within each group the gases are classified according to abundance in the atmosphere.

2.3
Gases Selected

More than 200 chlorinated gases have been identified in air (see Greadel 1979; Greadel et al. 1986). Of these gases, many are so short-lived and are produced in sufficiently low quantities that no appreciable concentrations are expected in most of the atmosphere. Such gases are present in localized environments, inside buildings and perhaps in or near large industrial complexes and cities. The importance of measurements of these gases is mostly to estimate exposure of workers and residents of nearby areas, as some of the gases are carcinogenic or could create health problems with sufficient exposure. Urban and indoor measurements have also been reported for many gases that are important for global atmospheric chemistry of reactive chlorine. We have not made use of these data as it is not possible to simply relate the results to the concentrations in the remote atmosphere. It is noteworthy that the measured high concentrations of these chlorinated gases is clear evidence of their anthropogenic sources (Lillian et al. 1975; Ciccioli et al. 1976; Pellizzari et al. 1976; Singh et al. 1977c, 1981, 1982; Pellizzari 1982; Harkov et al. 1983).

The criterion for selecting gases to be considered here was the expectation of measurable concentrations in the non-urban troposphere of a few parts per trillion or more. Gases with smaller concentrations do not contribute significantly to the atmospheric burden of chlorine. If emitted from the surface, as from industrial activities, gases with concentrations of a few parts per trillion or less are not expected to be significant in ozone depletion. In many cases the instruments used to measure the gases of environmental or scientific interest also result in the measurement of other gases that may be of marginal importance or exist at very low concentrations. Data on some such gases have been published and will be discussed briefly here.

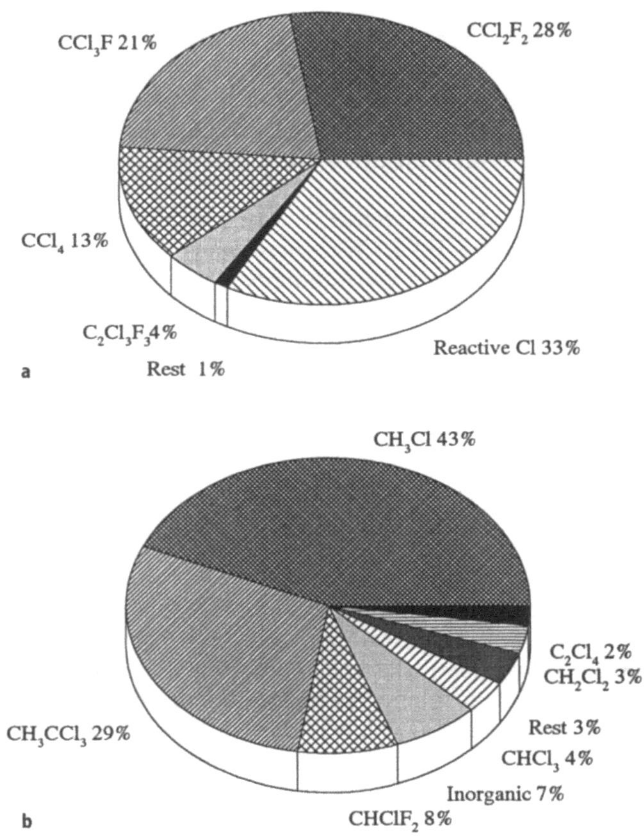

Fig. 1. **a** The amount of tropospheric chlorine from long-lived chloroflurocarbons (in 1995) and the total tropospheric chlorine from the reactive chlorine-containing gases. **b** The contribution of gases (as Cl) to the total reactive chlorine concentration in the troposphere. There are about 23 Tg of chlorine in the atmosphere, with about 14.8 Tg as long-lived gases, 2.7 Tg as moderately reactive gases and 5.7 Tg as reactive gases

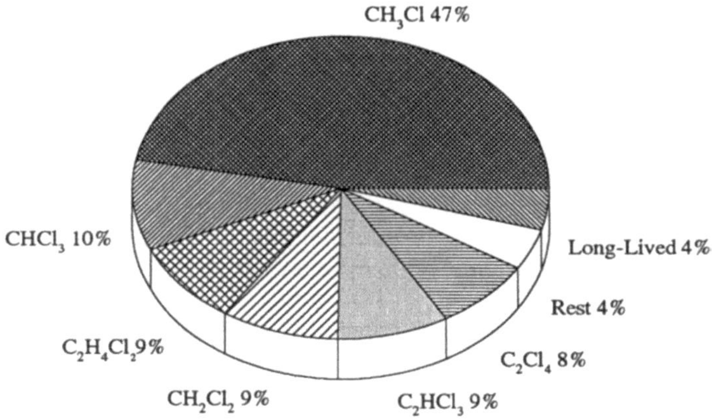

Fig. 2. Emissions of reactive chlorine-containing gases. For comparison the emissions of long-lived chlorofluorocabons are included (for 1995). The emissions of the CFCs in 1995 are, however, much lower than in previous years because of the scheduled phase out of these gases under the Montreal Protocol

Figure 1 is a summary of the gases that constitute the chlorine budget and their relative abundances in the atmosphere. In Fig. 1a, we see that reactive chlorine constitutes about a third of all the gaseous chlorine present in the atmosphere. The reactive chlorine component is dominated by methyl chloride, which is close to half, while methylchloroform makes up a significant part of the remainder (Fig. 1b). The other gases together make up about 27%, of which about 7% is inorganic chlorine, here taken to be hydrogen chloride. The emissions of reactive chlorine into the atmosphere are also dominated by methyl chloride at about 47%, followed by chloroform at 10% (see Fig. 2). The long-lived gases, which dominate the atmospheric concentration of gaseous chlorine, represent a small source, as would be expected because of their long atmospheric lifetimes (from Eq. 2). The emissions are particularly low at this time because of the effects of the Montreal Protocol. The Montreal Protocol and its amendments are international agreements to phase out the production of ozone-depleting compounds by the year 2000. The implementation of this program has resulted in very low emission rates at this time for most chloroflurocarbons and other gases involved in ozone depletion. The amounts of the long-lived gases that are present in the atmosphere came about from many years of emissions that were 5–10 times more than the current emission rates. So the concentrations at present are much higher than the current emissions would suggest. With this overview in mind, we can discuss what is currently known about the individual gases that constitute the reactive chlorine budget.

3 Reactive Chlorine-Containing Compounds (Lifetimes of < 1 yr)

3.1 Chloromethane (Methyl Chloride, CH_3Cl)

There are some 5 Tg of methyl chloride in the earth's atmosphere, making it one of the largest reservoirs of gaseous chlorine. It is the most abundant chlorine-containing gas in the atmosphere with average concentrations of about 600 pptv in the troposphere.

Systematic global measurements, taken over 15 years, show no major trends, except for a decline of concentrations in the middle 1980s and steady values since. The concentrations vary cyclically during the year. Highest values in the northern hemisphere are observed between February and April of each year, and the lowest values between August and October. Southern hemisphere cycles are approximately the reverse. The difference in the average concentrations during these periods is taken as the amplitude of the seasonal cycle. In the northern polar atmosphere the cycle amplitude is 16 pptv and goes down to about 10 pptv in the middle and tropical northern latitudes. It is smaller still, at about -3 pptv, in the southern tropics, and -5 pptv at the middle and high southern latitudes. This cycle is mostly driven by the seasonal nature of tropospheric OH radicals, which remove methyl chloride from the atmosphere (Khalil and Rasmussen 1997a).

The annually averaged spatial distribution of methyl chloride is characterized by higher concentrations in the tropics and lower concentrations at middle and high latitudes (Rasmussen et al. 1980; Khalil and Rasmussen 1997a). During some seasons the latitudinal distribution is more uniform and the tropical bulge is obscured by the seasonal cycles. Measurements taken from ship cruises or aircraft flights may not capture the long-term tropical high concentrations because of these seasonal variations. Nonetheless, other measurements spanning many latitudes either show higher tropical concentrations, or are compatible with the results from the long-term studies within the observed variability and uncertainty (Singh et al. 1979, 1983; Koppmann et al. 1993; Atlas et al. 1993). In the long term annual averages, no significant difference in concentration is observed between the two hemispheres. These characteristics - lack of a trend and inter-hemispheric gradient - imply that the major sources of CH_3Cl cannot be from industrial activities or consumer goods, and that many sources are likely to be natural, although what the natural sources are and how much these sources emit is still not accurately known.

Methyl chloride reacts with hydroxyl radicals (OH) giving it a lifetime of about 1.3 years (Khalil and Rasmussen 1997a; WMO 1995). Some CH_3Cl is lost in the stratosphere, where it reacts with stratospheric OH radicals and is photodissociated by solar ultra-violet radiation. There is also evidence that some soils may consume methyl chloride (Khalil and Rasmussen 1997b). There may be yet other loss processes, such as reactions with chlorine radicals in the marine boundary layer, that are difficult to quantify at present. Of these loss processes, reaction with OH is the most significant, and it is unlikely that any unquantified processes would be of comparable magnitude. The combined effect of these processes is an annual loss of about 3.5 Tg/yr. Since there is no systematic trend, the total annualized emissions must also be the same according to the mass balance discussed earlier (Eqs. 1 and 2). Because of the observed latitudinal distribution of the concentrations, and more so because of the high tropospheric OH radicals in the tropics, some 80% of this 3.5 Tg/yr must come from the tropics.

The source of methyl chloride from combustion processes has been know for a long time. In some parts of the United States, particularly in the Pacific northwest, high wintertime concentrations have been observed in urban and suburban areas. These are associated with wood burning in fire places and wood stoves for heating homes. In such environments there are often no other sources of CH_3Cl during the winters, and atmospheric oxidation is so slow as to make methyl chloride an inert tracer of wood-burning pollution. This has been used to estimate the amount of atmospheric fine particles that are contributed by wood burning in these environments (Khalil et al. 1983; Edgerton et al. 1986). Typical concentrations in polluted environments, with wood burning as a major source of methyl chloride, are 670 – 800 pptv, compared to background levels at these latitudes of 600–650 pptv during winters. The global emissions from biomass burning are estimated to be about 0.5 Tg/yr (Crutzen et al. 1979; Andrea 1991; Rudolph et al. 1995). Measurements taken in Beijing in 1985 and 1986 showed typical monthly average concentrations of 720–1250 pptv, which are substantially higher than background levels, of around 600 pptv, at similar latitudes. Low-temperature coal burning may be the source of this excess, but on a global scale this source is likely to be small.

Certain fungi can also produce CH_3Cl in significant quantities. At present the global emissions from this source are not known with certainty but appear to be in the range of 0.1–0.3 Tg/yr (Cowan et al. 1973; White 1982a, b; Harper 1985; and David Harper's unpublished data 1997 communicated to the author).

Until recently, it was thought that oceans are the major source of atmospheric methyl chloride at 3–6 Tg/yr (Singh et al. 1979; 1983; IPCC 1995). The early measurements clearly demonstrated that the oceans were a major source of methyl chloride. Recent experiments have shown, however, that the oceans are not likely to account for more than 0.8 Tg/yr (0.5–0.8 Tg/yr) (Tait et al. 1994; Moore et al. 1996; Khalil and Rasmussen 1997a; and unpublished work of J. Lobert and others from NOAA/CMDL).

There are a number of small sources such as industrial processes and possibly coal burning. There are a few industrial uses of methyl chloride as a methylating agent for the manufacture of chemicals and other processes which amount to less than 0.1 Tg/yr (Aucott et al. 1997a). Emissions from the known sources described above add up to at most about 2 Tg/yr, leaving a sizable gap of unexplained emissions to balance the budget. Major emissions from the industrialized middle latitudes are inconsistent with the observed latitudinal distribution of concentrations. All indications are, therefore, that the major sources are in the tropics, both accounted for and missing from the present budget.

Methyl chloride may also be removed by hydrolysis or other processes in the oceans; this would cause net fluxes from the ocean to the atmosphere in some regions, and an uptake of CH_3Cl in other parts of the oceans. Observations show a net flux of methyl chloride from the atmosphere to the oceans in the cold waters of the higher latitudes in both hemispheres (Tait et al. 1994; Moore et al. 1996; Khalil et al. 1997a; J. Lobert, unpublished work). But at most middle and tropical latitudes there is a net flux from the oceans into the atmosphere according to the same studies, with an indication of higher emissions in the tropics. This observation suggests that the oceanic flux of methyl chloride may be quite variable from one location to another. It also leads to higher tropical emissions, not only because the flux is higher, but also because of the large area of the warm oceans in the tropical regions compared to higher latitudes. This tropical bulge in the oceanic emissions is compatible with the high tropical emissions required to balance the observed concentrations in the tropics, but it does not explain more than about 30% of the emissions reguired from this region to balance the observed atmospheric concentrations. The discrepancy in emissions can only be resolved by further research. Possible explanations for this gap are that the measured reaction rate constant with OH is too fast and therefore we have overestimated the global loss rate, or there may be unknown sources. Sources may include atmospheric formation of methyl chloride either by heterogeneous or gas-phase processes involving the chlorine emitted by the oceans – such processes have not been investigated (see Sect. 7 for a summary).

3.2 **Trichloromethane** (Chloroform; $CHCl_3$)

The average concentration of chloroform is about 18 pptv at the surface, and around 15 pptv as an average in the troposphere. It is nearly twice as abundant in the northern hemisphere compared to the southern hemisphere. Although there are few studies that can be compared with each other, there is fair agreement on atmospheric concentrations between the long-term data and studies from ship cruises (see Singh et al. 1983b; Class and Ballschmiter 1986; Atlas et al. 1993; Khalil and Rasmussen 1997c).

In the decadal time series, highest concentrations are observed at the middle northern hemisphere site. Systematic global measurements taken during the last decade show no trends (Khalil and Rasmussen 1997c; see Sect. 7), but the concentrations undergo seasonal variations that are likely to be driven by the seasonality of OH, with which it reacts rapidly. The annual and globally averaged concentrations are also variable, ranging from 22 to 16 pptv.

The lifetime of chloroform from reaction with OH radicals is about 0.5 year. The losses due to reactions with OH are around 500 Gg/yr with considerable uncertainty. The oceans are a source of around 360 Gg/yr (Khalil et al. 1983). Moreover, the soils are also a source of chloroform in the few studies currently available. The data are too few to obtain an accurate estimate, but emissions are estimated to be several 100 Gg/yr. Most of the industrial production of chloroform is used for manufacturing F-22 and would therefore not be released into the atmosphere (see e.g. NAS 1978). It is also used in the production of plastics and vinyl chloride. Emissions from other uses have been estimated by Aucott et al. (1997) to be about 60 Gg/yr. Of these, the largest source is pulp and paper manufacturing at about 30 Gg/yr, followed by water treatment at about 20 Gg/yr. Biomass burning may also release chloroform but, based on limited data, the source appears to be very small (Rudolph et al. 1995). These emission rates are so uncertain that it is difficult to say whether the budget is balanced or whether new sources and sinks are needed to account for the observed atmospheric concentrations. The removal is assumed to be mostly from reaction with OH radicals, but it is possible that there are photochemical or other atmospheric removal processes as well.

The large emissions required in the higher latitudes cannot be explained by an oceanic source. The distributions of the soil emissions and industrial sources would be compatible with the expected latitudinal emissions, but these would have to be quite large. It is possible, therefore, to obtain a budget that is compatible with atmospheric observations by adopting a sizable terrestrial source concentrated in the middle northern latitudes and a smaller oceanic source to balance the budget at other latitudes (See notes to Table 1 in Sect. 7 for more information).

3.3 Dichloromethane (CH_2Cl_2)

The concentrations of dichloromethane, or methylene chloride, are about 25 pptv on average. It is mainly removed by OH reactions. This leads to estimated emissions of some 0.6 Tg/yr, which is comparable in magnitude to the estimated emissions of most of the other main reactive organic chlorine gases (other than CH_3Cl). The marine boundary layer latitudinal distribution shows high concentrations in middle northern latitudes of some 40 pptv and decreasing to about 17 pptv in the southern hemisphere, which suggests that there are sizable anthropogenic sources (Singh et al. 1983b; Koppmann et al. 1993; Khalil and Rasmussen 1993; Atlas et al. 1993). Dichloromethane is used extensively in commercial and industrial processes as well as in many common consumer products (NAS 1978; NSC 1997).

There is evidence that the oceans are supersaturated with dichloromethane, but at present it is difficult to estimate the oceanic emission rate. The inter-hemispheric ratio is about 1.7, whereas, if all the emissions were in the northern hemisphere, the expected ratio would be around 3, due to its relatively short lifetime. Roughly 25% of the emissions, or some 150 Gg/yr, should be from the southern hemisphere to explain the smaller than expected inter-hemispheric ratio. If all this source is oceanic, then an upper limit for the oceanic emissions would be about 300 Gg/yr, or roughly half the annual emissions. Dichloromethane is also released from biomass burning, perhaps around 70 Gg/yr (Rudolph et al. 1995). Estimates of the industrial production of dichloromethane in the U.S. alone for 1987 are 280 Gg (NSC 1997). How much is emitted into the atmosphere is not accurately known, but enough is produced to account for the total budget along with possible natural emissions.

3.4 Phosgene ($COCl_2$)

Phosgene is not directly emitted into the atmosphere but is a byproduct of the oxidation of other less reactive chlorine-containing gases. It is, however, a sizable reservoir of chlorine in the atmosphere. It is particularly important because it represents the continued presence of reactive chlorine in the atmosphere even after the parent compounds have been removed. This active chlorine can be carried to the stratosphere. In the troposphere it is produced by the oxidation of C_2Cl_4, CH_3CCl_3, $CHCl_3$, and C_2HCl_3, with yields of 0.4–1 molecules of phosgene per molecule of precursor oxidized, as described by Kinder et al. (1995). A smaller yield of about 10% is estimated for the oxidation of dichloromethane (Helas and Wilson 1992). In all, some 0.9 Tg Cl/yr are produced (Helas and Wilson 1995; Kinder et al. 1995). Most of it is removed by hydrolysis in clouds and subsequent rain and deposition to the oceans. Atmospheric concentrations have been measured to be 15–50 pptv, with prevailing tropospheric values of around 20 pptv (Singh 1976; Wilson et al. 1988). The mass balance gives a lifetime of 60–80 days.

3.5 Chlorinated Ethylenes (C_2HCl_3, C_2Cl_4)

Trichloroethylene (TCE) and perchloroethylene (PCE) are used mostly as solvents. Their atmospheric lifetimes, however, are quite different, although both are less than a year, mainly due to reactions with OH radicals. Trichloroethylene has a lifetime of a few days (about 6 on average), and perchloroethylene one of about 150 days. This is because it does not have a carbon-hydrogen bond that generally leads to the most rapid reactivity with OH. The lifetime of C_2Cl_4 may also be affected by the reactions with Cl radicals in the marine boundary layer, but at present the concentrations of Cl radicals are not known (Singh et al. 1996).

The concentrations have been measured in several independent studies. The earlier publications by Singh et al. (1977, 1979, 1983b) and Rasmussen and Khalil (1982) show significantly higher concentrations than the more recent studies reported by Koppmann et al. (1993), Wiedmann et al. (1994), Wang et al. (1995), and Youkouchi et al. (1996). The latitudinal patterns are similar in these studies, suggesting that some of the disagreement is from differences of absolute calibration. Because of the short lifetime, substantial seasonal cycles are expected, although observational data are not available for C_2HCl_3. Seasonal cycles are, therefore, also a source of differences between the concentrations reported from various studies. The high values reported earlier have been adjusted to the scale reported in Koppmann et al. (1993) for C_2HCl_3 and to the scale of Wang et al. (1995) for C_2Cl_4 as these are likely to be more accurate. With these adjustments, the mean concentrations of trichloroethylene is about 3 and 0.7 pptv in the northern and southern hemispheres, respectively, and concentrations of perchloroethylene are about 17 and 3 pptv, respectively, in the northern and southern hemispheres (Singh et al. 1983b; Koppmann et al. 1993; Wang et al. 1995; Khalil and Rasmussen 1997b). The large inter-hemispheric ratio is consistent with the idea that these gases are mostly from human activities (see note 1 e, Table 1, Sect. 7).

Although the concentrations in the southern hemisphere are quite small, some natural emissions can be accommodated. Measurements have shown that most of the oceans are supersaturated with these gases, implying a natural source. There are very few data currently available to estimate how much perchloroethylene and trichloroethylene might be emitted by the oceans (Khalil and Rasmussen 1997b). Supersaturations for PCE have also been observed by J. Lobert from NOAA/CMDL and V. Karapalov (unpublished results communicated to the author). The data we have suggest a source of 20 Gg/yr for perchloroethylene and 30 Gg/yr for trichloroethylene (see Notes to Table 1, Sect. 7). The range of emissions compatible with the data are between 20–100 Gg/yr for either compound. Whether there are other natural sources is not known. The remaining amounts are likely to be emitted mostly from manufacturing losses, commercial uses and consumer products. Perchloroethylene is widely used for dry cleaning, as a degreasing solvent and in the manufacture of F-113. It is also used in inks, adhesives, sealants, polishes, lubricants, and pesticides. Trichloroethylene is used primarily as a degreasing solvent, mostly in the automotive and metals industries. It is also used in many other industrial processes and household products. The industrial production can account for the observed concentrations, but a rigorous mass balance has not been carried out, and the actual amounts released to the atmosphere have not been established (NSC 1997).

3.6
Chlorinated Ethanes ($C_2H_4Cl_2$, $C_2H_2Cl_4$)

The chloroethanes have been measured in a few experiments. The observed concentrations are small at about 6 pptv for dichloroethane ($C_2H_4Cl_2$) and around 0.1 pptv for tetrachloroethane ($C_2H_2Cl_4$) (Class and Ballschmiter 1987; Khalil and Rasmussen 1997b). The inter-hemispheric ratios are high at 10 for $C_2H_4Cl_2$ and 5 for $C_2H_2Cl_4$ suggesting mostly anthropogenic sources. For $C_2H_4Cl_2$ concentrations in the oceans are in equilibrium with the atmosphere. There are no other data to test for the existence of natural emissions.

Based on the reactions with OH radicals, the annual losses of $C_2H_4Cl_2$ are about 0.7 Tg/yr (or 0.5 Tg Cl/yr) and for $C_2H_2Cl_4$ the losses are less than 0.01 Tg/yr. These calculations suggest a negligible source of reactive chlorine to the atmosphere from $C_2H_2Cl_4$, but on the other hand there could be a sizable anthropogenic source of $C_2H_4Cl_2$. These findings warrant further study to quantify the importance of $C_2H_4Cl_2$ in the tropospheric chlorine budget.

Dichloroethane ($C_2H_4Cl_2$) is used to make vinyl chloride, which has many uses. It is used in household applications including glues, cleaning agents, and paint removers; as an agricultural fumigant; and as a solvent for industrial applications. The 1990 U.S. production of dichloroethane was several Tg, most of which was probably used for manufacturing other chemicals, but the losses during processing, and from its other uses, could be sufficient to account for the estimated releases of 0.7 Tg/yr, making it an important contributor to the global reactive chlorine budget (NSC 1997). In contrast, tetrachloroethane ($C_2H_2Cl_4$) is produced in very small quantities, which is consistent with the low concentrations observed in the atmosphere. It is used for the manufacture of TCE, PCE, and other chemicals and as a solvent for special applications (NSC 1997).

3.7 Rare Chlorinated Organic Compounds

The existence of many other chlorinated gases has been documented, but in most cases the concentrations in background air are very small. Class and Ballschmiter (1986, 1987) report measurements of hexachlorobutadiene (C_4Cl_6) with average concentrations of about 0.3 and 0.1 pptv in the northern and southern hemisphere, respectively. Similarly low concentrations were also reported by Atlas et al. (1993). The lifetime of C_4Cl_6, based on its inter-hemispheric ratio (and assuming all emissions are in the northern hemisphere), is less than a year.

There are a number of other gases in this group with lifetimes of a few days to a month, including the chlorobenzenes, vinyl chloride (C_2H_3Cl), and vinylidene chloride ($C_2H_2Cl_2$), that have been observed in polluted air (Singh et al. 1975, 1981, 1982). Yokouchi et al. (1996) found an average of 0.05 pptv of $C_2H_2Cl_2$ in the arctic troposphere, which represents a negligible concentration in the atmosphere. They also reported very low levels of dibromochloromethane ($CHClBr_2$) averaging around 0.2 pptv. Atlas et al. (1993) report similarly low values of $CHClBr_2$ of between 0.1 and 0.4 pptv. Measurements of bromochloromethane (CH_2ClBr) from ship cruises showed an average over the oceans between 1–3 pptv (Khalil and Rasmussen 1997b), but the arctic measurements by Yokouchi et al. (1996) do not exceed 0.5 pptv.

3.8 Inorganic Chlorine Gases

Although most of the gases considered here are organic compounds, there are inorganic gases that play a significant role in the global budget and cycles of reactive chlorine in the atmosphere, and particularly in the marine boundary layer.

Hydrogen chloride is perhaps the most abundant inorganic chlorine-containing gas in the atmosphere with emissions of some 100 Tg Cl/yr (possibly much larger). The atmospheric concentrations, however, are not large as the lifetime is short at only about 1–2 days. Graedel and Keene (1995) have compiled existing data on the concentrations and fluxes of HCl. Measured concentrations are extremely variable and there are not many data from which to determine the global distribution. The best estimate of prevailing concentrations given by Graedel and Keene (1995) is about 200 pptv in the marine boundary layer based on many sporadic studies, with low concentrations of 50 – 100 pptv above about 2 km based on results from Vierkorn-Rudolph et al. (1984). Concentrations over continental areas are expected to be lower.

In regions near sources, concentrations can be an order of magnitude higher. Typical urban levels are found to be around 1.5–2.0 ppbv. Because the lifetime is so short, these high concentrations are not carried far into the remote atmosphere. Rather, the concentrations observed in the remote atmosphere must mostly be generated by local sources.

In the stratosphere, HCl concentrations are 1–3 ppbv (WMO 1985; Warneck. 1988) and increase with increasing altitude. The average concentrations is about 1 ppbv, making it the largest reservoir of chlorine in the stratosphere. Much of this HCl is expected to come from the Cl atoms liberated from the long-lived and moderately reactive gases considered here, as well as from CH_3Cl. As such, in contrast to tropospheric HCl, stratospheric concentrations are mostly from anthropogenic precursors (Froidevaux et al. 1985; Brasseur et al. 1990). It is a relatively inactive form of chlorine in the stratosphere and is transported down to the troposphere, but it contributes very little to the tropospheric budget.

Over the oceans, HCl can be formed by volatilization from sea salt aerosol by acid displacement reactions involving HNO_3 and H_2SO_4. Graedel and Keene (1995) estimate a source of 40–70 Tg Cl/yr would be required to balance the observed concentrations that average about 200 pptv in the marine boundary layer, with a lifetime, based on deposition processes, of 1–2 days. They show that this amount could be accounted for by acid displacement reactions. Other estimates of emissions from volatilization of sea salt have ranged from 200–1200 Tg Cl/yr (Cadle 1980; Cicerone 1981; Moller 1990). Volcanoes are estimated to emit <1–10 Tg/yr (Symonds et al. 1988; Cadle 1980) and anthropogenic activities could account for a few Tg/yr (Cicerone 1981; Graedel and Keene 1995). The loss of HCl due to reactions with OH could be 5–20 Tg/yr depending on the concentrations in the free troposphere. This amounts to 50–100 Tg Cl/yr as HCl that must be emitted to the atmosphere to balance these removal processes, but this number is still very uncertain. It is, however, very large compared to the emissions of organic chlorine-containing gases.

In addition to the sea salt dechlorination source, there are other possible pathways for the production of HCl in the marine boundary layer from photochemically active forms of inorganic chlorine such as Cl_2, HOCl, BrCl and $ClNO_2$. Such processes have been suggested by Keene et al. (1990, 1996) and Graedel and Keene (1995). They propose a chain of reactions starting with volatilization of Cl_2 from sea salt aerosol. The Cl_2 is split by photolysis to produce two chlorine radicals. The Cl radicals react, with lifetimes of few tenths of a minute, with the many hydrocarbons and other gases present in the atmosphere, resulting in the production of HCl. The HCl is scavenged by the sea salt aerosol. The combination of Cl_2 and HOCl (which is a product of chlorine chemistry in the marine boundary layer) range from <30–250 pptv (Pszenny et al. 1993). Based on these concentration data, the photolysis rates for HOCl and Cl_2 and an assumed range of the ratio of Cl_2/HOCl between 0 and 1, Graedel and Keene (1995) estimate that <100–800 Tg Cl/yr as HCl could be cycled through this process. Addition of this process results in much higher levels of Cl radicals compared to the case when the Cl_2+HOCl cycle is not included, yet it would remain consistent with the observed concentrations of HCl (which would have a shorter lifetime than previously estimated because of the addition of the sea salt scavenging sink). The other reactive gases mentioned would also undergo the same process resulting in the production of Cl radicals.

The atmospheric abundance of Cl radicals depends on the production from the reaction of HCl with OH and the photochemical processes discussed above. Atomic chlorine, if its concentrations are high enough, can be an important contributor to the removal of many man-made and natural trace gases, such as methane, which react rapidly with these radicals (see Keene et al. 1996). Because the processes that produce Cl radicals are not understood at present, there is no consensus as to the concentration of Cl radicals in the marine boundary layer with estimates ranging from negligible ($< 10^3$ molecules/cm^3) to 10^6 molecules/cm^3. In some studies indirect methods have been used to estimate the concentrations of Cl radicals, but the uncertainties in the estimates are quite large and do not resolve the question of whether there are enough Cl radicals in the marine boundary layer to add significantly to the removal of environmentally important trace gases (Singh and Kasting 1988; Keene et al. 1990, 1996; Pszenny et al. 1993; Jobson et al. 1994; Singh et al. 1996a, b; Rudolph et al. 1996; Laux et al. 1996; Sander and Crutzen 1996; Vogt et al; 1996; Wingetner et al. 1996).

4
Moderately Reactive Compounds (Lifetimes of 1–10 Years)

4.1
1,1,1-Trichloroethane (Methylchloroform; CH_3CCl_3)

Methylchloroform is a degreasing solvent that has taken over the markets previously occupied by trichlorethylene and perchloroethylene. It is much less reactive than the compounds it has replaced and a lesser health risk in the workplace. Detailed and systematic global measurements of methylchloroform have been taken since the early 1980s. One of the scientific interests in the compound is its use as a tracer for estimating tropospheric OH concentrations. Since its industrial production and uses are well documented, it is possible to accurately estimate its annual emissions. Knowing the emissions and the atmospheric concentrations, the mass balance equation (Eq 2) can be solved for the third component – namely the annual loss rate, which is directly related to the average OH concentrations (Lovelock 1997; Singh 1977a; Khalil and Rasmussen 1984; Prinn et al. 1992; 1995). The loss rate is K [OH] C, where K is the reaction rate constant for $OH+CH_3CCl_3$ which is measured in laboratory studies, and C is the concentration of methylchloroform, which is known from atmospheric measurements. The OH concentration can therefore be calculated. Using this method, with the current knowledge of its reaction rate constant and other sinks, the average OH concentrations are calculated to be about 10^6 molecules/cm^3, which are in good agreement with most photochemical models of OH. It may even be possible to detect long-term trends of OH concentrations using a long time series of methylchloroform measurements. While methylchloroform provides a powerful tool for the study of the oxidizing capacity of the atmosphere, it is not possible to obtain much information about the latitudinal, vertical or seasonal changes of OH. This role is filled by photochemical models.

The latitudinal distribution of methylchloroform was characterized by about 4% higher concentration in the northern hemisphere compared to the southern hemisphere in 1995. This inter-hemispheric ratio was 20% at its peak in the 1980s and early 1990s.

Methylchloroform concentrations have changed dramatically over the last two decades. In the early 1980s the concentrations increased at rapid rates of around 4–5%/yr as the compound filled its new markets and applications. Because of the Montreal Protocol, methylchloroform production is scheduled to be phased out. In recent years production has dropped from 270 Gg/yr in the 1980s to about 126 Gg/yr in 1993 (Midgley and McCulloch 1995). Because of its modest lifetime, the atmospheric concentrations of methylchloroform are already declining rapidly. Current trends are about –16 pptv/yr or –12.5%/yr. Methylchloroform trends will have a significant effect on the future concentrations of reactive chlorine in the atmosphere (troposphere and stratosphere) as it represents about 2 Tg Cl in the atmosphere out of about 6.5 Tg of organic chlorine (See Table 1, Sect. 7).

4.2 **Chlorodifluoromethane** ($CHClF_2$; F-22)

Chlorodifluorormethane belongs to a class of compounds referred to as hydrochlorofluororcarbons (HCFCs). These compounds are safer for the ozone layer since they react with tropospheric OH radicals and often have fewer chlorine atoms. The lifetime of F-22 is around 11 years due to reactions with OH and about 9 years when other removal processes, such as the stratosphere, are taken into account.

Present (1995) concentrations of F-22 are about 117 pptv, with an inter-hemispheric gradient of about 15 pptv (Montzka et al. 1993, 1996). Most of the emissions are in the northern hemisphere, and the global distribution is in agreement with this fact. The concentrations are increasing at a relatively rapid rate of about 6 pptv/yr or 7%/yr (Montzka et al. 1993). There is a slow but steady growth in production and emissions (AFEAS 1997). It is likely, therefore, that concentrations will continue to rise as F-22 fills some of the markets that were previously occupied by the chlorofluorocarbons F-11 and F-12, particularly in air conditioning, refrigeration and foam blowing. No natural sources are known, but potential sinks, such as oceans and soils, may exist.

4.3 **Dichlorofluoromethane** ($CHCl_2F$; F-21)

The concentrations reported by Penkett et al. (1980) range from 0.1–2.5 pptv in the troposphere resulting in a global average concentration of 1–2 pptv. Further measurements by Crescentini et al. (1986) showed concentrations mostly in the range of 0.1–1 pptv. Other results published by Singh et al. (1977b) suggested average concentrations of around 5 pptv, but the authors were uncertain of the reliability of the data. Most studies also reported occasional high values. Although a budget of F-21 is not possible with such limited and variable data, it is apparent that the concentrations are low and not likely to contribute significantly to the reactive chlorine budget.

4.4 Other Chlorinated Gases

There are a few other moderately reactive chlorinated gases that have been identified in the atmosphere. Class and Ballschmiter (1986, 1987) have reported a latitudinal distribution of hexachloroethane (C_2Cl_6). The concentrations observed were approximately 0.3 pptv and there is little difference between the hemispheres (see also Singh et al. 1979).

$CClF_3$ (F-13) has a concentration of about 3 pptv with a small inter-hemispheric gradient (Penkett et al. 1981). Because of its long lifetime estimated at some 600 years (WMO 1995, chap. 13), the annual emissions are likely to be negligible for the global chlorine cycle.

5 Long-Lived Chlorinated Compounds (CCl_2F_2, CCl_3F, CCl_4, $C_2Cl_3F_3$, $C_2Cl_2F_4$, C_2ClF_5, $CClF_3$)

The long-lived chlorinated compounds are all anthropogenic and have no known reactivity in the troposphere. There may be minor removal pathways in the oceans and the soils for some of these gases (Khalil and Rasmussen 1989; Denovan and Strand 1992). Most of the removal of these gases takes place in the stratosphere by photo-dissociation by ultra-violet radiation. As such these gases are not regarded as part of the "reactive chlorine" group. These are included here mainly to provide a broader context for the reactive chlorine gases and to show the relationship between the budgets of the reactive and non-reactive chlorine-containing gases.

It is noteworthy that this group represents significantly more chlorine in the atmosphere than the previous two groups of gases combined. This dominance is even more significant in the stratosphere. Concentrations of reactive chlorine gases tend to decrease faster in the stratosphere than the long-lived gases, partly because of the high concentrations of OH in the stratosphere and other removal processes that exist in addition to photo-dissociation. Even in the troposphere, the long-lived gases have far exceeded natural chlorine concentrations. Presently, these gases constitute some 2.6 ppbv of chlorine in the troposphere, while the reactive and moderately reactive chlorine gases are together about 1.3 ppbv (see Table 1, Sect. 7).

This dominance of the long-lived gases in the total chlorine budget is a consequence of their long lifetimes which allow for a buildup over many years of emissions. In the chlorine cycle, it appears that naturally produced gases are short-lived. As will be discussed later, all the known natural gaseous chlorine is in the "reactive" group. In this group gases have known oceanic, soil and microbial sources, although none is entirely natural. But in the groups of long-lived and moderately lived gases there is no established evidence of any natural sources.

In the past, these gases have served important industrial and consumer needs. For the 20-year period between 1975 and 1995, about 6.4 Tg of F-12 and 5.3 Tg of F-11 were produced for commercial use. During this time some 30% of

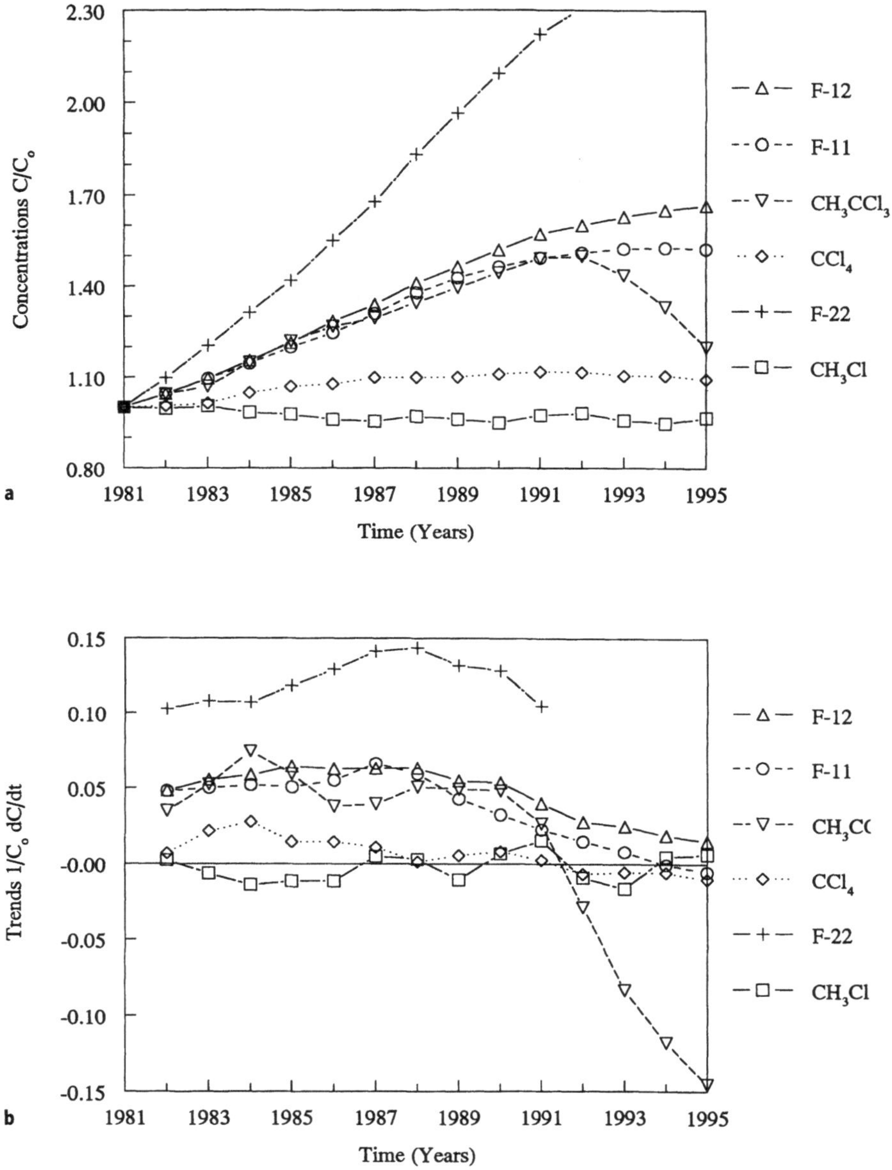

Fig. 3. a The relative concentrations of major chlorine-containing gases in the atmosphere. The actual concentrations can be recovered from the graph by multiplying by the concentration (C_o) in the base year (see text). **b** The trends of chlorine-containing gases. Actual trends can be recovered by multiplying by Co as in **a** above. Concentrations of CH_3CCl_3, F-11, CCl_4, and F-113 (not shown here) are falling, while the trends of F-12 are much slower than in earlier years. Trends of CH_3Cl and $CHCl_3$ (not shown here) are near zero and insignificant

the production was used for aerosol propellent, 3% of F-12 and 17% of F-11 were used for blowing open cell foams, 7% and 39% of F-12 and F-11, respectively, were used for blowing closed cell foams and 54% and 9%, respectively, were used for refrigeration and air conditioning (AFEAS 1997). The amounts used in refrigeration and foams are still not released into the atmosphere and will be a source of these compounds for many years to come (Khalil and Rasmussen 1993). In recent years, the industrial production of CCl_4 was used mainly for manufacturing F-11 and F-12. The cumulative production between 1958 and 1985 is estimated to be 3.9 Tg (Simmonds et al. 1988). F-113, F-114 and F-115 have been used for aerosol applications, dry cleaning, blowing foams, and refrigeration and in the electronics and other industries for specialized applications (AFEAS 1997).

The recent AFEAS (1997) estimates of annual emissions show dramatic cuts in production because of the Montreal Protocol. The industrial production of F-11 in 1995 was about 8% of the peak production in 1988; for F-12 it was about 23% of its peak value in 1988 and for F-113 it was less than 10% of peak production levels. The rates of increase of these gases in the atmosphere, therefore, are either slowing down or have already reversed and concentrations are falling. In the next several decades, these concentrations are expected to decline considerably (see Table 1, Sect. 7).

The trends of the main chlorine-containing gases are shown in Fig. 3. To compare the changes in concentrations that have taken place over the last fourteen years, we have plotted the ratio (C/C_0) of the globally and annually averaged measured concentration [C(t)] and the concentration during 1981 (C_0). In the companion figure we calculate the trends as $1/C_0\ dC/dt$. The results show that the atmospheric trends of the CFCs and CH_3CCl_3 are all decreasing. The trends of F-11, CCl_4 and CH_3CCl_3 are negative in the most recent years showing a decreasing concentration in the atmosphere. Although F-12 trends are now quite small, the concentrations continue to increase. For F-22, there a continuing growth as expected from its current applications. The actual annual concentrations and trends can be recovered from this figure by multiplying the values by C_0, which are 306, 181, 105, 115, 44, 619 pptv for F-12, F-11, CH_3CCl_3, CCl_4, F-22 and CH_3Cl, respectively.

The main uses of these chlorofluorocarbons will be filled by hydrochlorofluorocarbons (HCFCs) and hydrofluorocarbons (HFCs). Specifically, HCFC-143b, HCFC-141b, and HCFC-134a will be used for blowing foams, as solvents, and for refrigeration applications, respectively. These new compounds are discussed next.

6 New Hydrochlorofluorocarbons

A large number of new chemicals are beginning to be manufactured and distributed to replace the well-known chlorofluororocarbons (CCl_2F_2 and CCl_3F) and other fully halogenated methanes and ethanes (e.g. F-113, F-114, F-115), which are being phased out. Most of the HCFC replacement compounds have very few chlorine atoms and have carbon–hydrogen bonds that make them reactive with tropospheric OH radicals. Nonetheless, most of the HCFCs have atmospheric lifetimes due to OH reactions that are 1–20 years (see Table 1 in Sect. 7 and note 6).

There are two classes of replacement compounds: the hydrochlorofluorocarbons (HCFCs) and hydrofluororcarbons (HFCs). The most important HCFCs are HCFC-22, HCFC-142b, and HCFC-141b, as these are already in production. HCFC-22, discussed earlier, has been used for air conditioning and related applications for some time. The other HCFCs have been produced only during the last several years. The first estimates of the production and release of HCFC-141b and HCFC-142b have been published by AFEAS (1997). Measurements show concentrations around 3 pptv for HCFC-141b and 7 pptv for HCFC-142b. These are consistent with the published estimates of anthropogenic emissions of HCFC-142b but are higher by about a factor of two compared to the concentrations expected for HCFC-141b from estimated industrial emissions (Montzka et al. 1996; Elkins et al. 1996; AFEAS 1997).

The ozone-depletion potentials (ODP) and global-warming potentials (GWP) for the replacement compounds have also been calculated (WMO 1995; IPCC 1994; Kindler et al. 1995). These compounds are 10 to 50 times less effective than CFC-11 for destroying the ozone layer (ODPs range from 0.02–0.1). The global-warming potentials, however, continue to be large compared to CO_2 with values of 300–4000, but these potentials are significantly less than CFCs 11 and 12. Since the concentrations of these gases are expected to remain low, the actual effect on the earth's climate is likely to be negligible.

The HCFCs are promising candidates for estimates of tropospheric OH concentrations in the same way as methylchloroform was used in the past. These calculations will require accurate and precise atmospheric measurements. This will take some time because the concentrations are very small at present and the measurement technology is still being developed.

7 The Reactive Chlorine Budget

The global budgets of the chlorine-containing gases discussed in this chapter are summarized in Table 1. This section includes general and detailed notes documenting the data and calculations presented in the table. The notes that follow are numbered according to the numbers representing the columns of the Table.

(1) **Concentrations.** The concentrations are given in four equal area regions at the surface of the earth, Cn for northern middle and high latitudes (30 N–90 N), Cnt for northern tropics (0 N–30 N), Cst for the southern tropics (0 S–30 S) and Cs for the middle and high southern latitudes (30 S–90 S). The four divisions allow latitudinal data to be cited for most gases of interest. For gases with lifetimes longer than a few months, the latitudinal changes of concentration are the most significant features of the global distribution as vertical mixing ratios tend to be relatively constant (except in the stratosphere). The latitudinal changes of concentration are a qualitative reflection of the distribution of sources and sinks.

Table 1. Chlorine-containing gases in the atmosphere arranged by reactivity and tropospheric abundance from highest to lowest

Gas	Concentration								Lifetime	Emissions	Global Effects	
	Troposphere							Stratosphere			ODP	GWP
	Cn	Cnt	Cst	Cs	Trend	Cave	Natural					
	pptv				pptv/yr	pptv Cl	pptv Cl	pptv Cl	Yrs	Tg Cl/yr		
	(1)				(2)	(3)	(4)	(5)	(6)	(7)	(8)	20 yr Horizon
Reactive (≤ 1 yr)												
CH_3Cl	585	607	605	585	~ 0 ± 1.0	554	469	290	1.3	2.6	0.02	27
HCl	200	200	200	200	+	90	90	1000	0.004	90	nk	nk
$CHCl_3$	29	18	12	15	~ 0 ± 0.5	47	22	25	0.5	0.5	0.008	15
CH_2Cl_2	41	19	17	17	nk	39	19	23	0.4	0.5	< 0.001	28
$COCl_2$	18	19	15	11	nk	32	8	20	0.2	0.9	0.003	nk
C_2Cl_4	17	10	4	4	nk	27	3	18	0.3	0.4	0.006	nk
$C_2H_4Cl_2$	16	4	1	1	nk	8	0	5	0.1	0.5	nk	nk
C_2HCl_3	3.8	2	0.7	0.7	nk	1.6	0.6	< 1	0.02	0.5	< 0.001	nk
CH_2BrCl	2.6	1.5	1.6	1.1	nk	1.4	0.9	< 1	0.4	0.01	nk	nk
Sub-Total					nk							
Chlorine (ppbv)	0.89	0.81	0.76	0.74	nk	0.80	0.6	1.4	–	–	–	–
Chlorine (Tg or Tg/yr)	1.13	1.03	0.96	0.94	nk	4.1	3.1	1.6	0.04	96	–	–
Organic Cl (Tg or Tg/yr)	1.02	0.91	0.85	0.82	~ 0	3.6	2.7	0.4	0.6	6.0	–	–
Moderately Reactive (1–10 yrs)												
CH_3CCl_3 (F-140a)	136	133	122	121	– 15.7 ± 0.6	373	0	177	5.2	0.12	0.12	360
$CHClF_2$ (F-22)	112	104	96	95	6.3 ± 0.3	102	0	71	11	0.09	0.05	4300
$CHCl_2F$ (F-21)	1.6	1.6	0.2	0.2	nk	1.7	0	< 1	1.7	0.01	nk	nk

Sub-Total					– 39.4							
Chlorine (ppbv)	0.51	0.49	0.45	0.45	–	0.5	0	0.25	–	–	–	
Chlorine (Tg or Tg/yr)	0.65	0.63	0.57	0.57	–	2.4	0	0.29	10.7	0.2	–	
Long Lived (> 10 yrs)												
CCl_2F_2 (F-12)	545	540	534	532	4.8 ± 0.3	1075	0	587	100	0.11	0.9	7900
CCl_3F (F-11)	279	277	275	273	– 1.1 ± 0.2	829	0	399	50	0.10	1	5000
CCl_4 (F-10)	127	126	125	124	– 1.1 ± 0.1	502	nk	230	40	0.02	1.2	2000
$C_2Cl_3F_3$ (F-113)	58	58	58	57	– 0.03 ± 0.1	174	0	94	90	0.02	0.9	5000
$C_2Cl_2F_4$ (F-114)	14	13	12	12	0.9	26	0	16	300	0.001	0.9	6900
C_2ClF_5 (F-115)	9	9	9	9	0.5	9	0	6	1700	0.003	0.4	6200
$CBrClF_2$ (H-1211)	3.2	3.0	2.9	3.0	0.11	3	0	2	20	0.002	5	nk
$CClF_3$ (F-13)	3	3	3	3	nk	3	nk	2	640	< 0.01	nk	8100
Sub-Total					4.2							
Chlorine (ppbv)	2.66	2.63	2.60	2.59	–	2.6	0	1.3	–	–	–	–
Chlorine (Tg or Tg/yr)	3.37	3.34	3.31	3.29	–	13.3	0	1.5	–	0.2	–	–
Total of all Categories					– 35							
Chlorine (ppbv)	4.1	3.9	3.8	3.8	–	3.9	0.6	3.0	–	–	–	–
Chlorine (Tg or Tg/yr)	5.2	5.0	4.8	4.8	–	19.8	3.1	3.4	–	96	–	–
Organic Cl (Tg or Tg/yr)	5.0	4.9	4.7	4.7	–	19.3	2.7	2.3	–	6.5	–	–
Replacement HCFCs												
$C_2H_3FCl_2$ (HCFC141B)	5	3.5	2.5	2.5	1.3	6	nk	nk	9	0.022	0.1	1800
$C_2H_3FCl_2$ (HCFC142B)	8	7	6	6	1	7	nk	nk	17	0.004	0.07	1200

Data from long-term sampling sites are from Barrow, Alaska (71.16°N, 156.5°W), Cape Meares, Oregon (45.5 N, 124 W), Mauna Loa and Cape Kumukahi, Hawaii (19.3°N, 154.5°W), Samoa (14.1°S, 170.6°W), Cape Grim, Tasmania (42 S, 145 E), and Palmer Station (64.46 S, 64.04 W) and the South Pole in Antarctica (90 S). The averages for the four regions are latitudinally weighted among these sites. For ship cruise data, averages are taken directly in each of the regions described above.

In some cases there were not enough data to estimate the concentrations in all four of the regions mentioned above. Such cases are noted below. For the moderately reactive and long-lived gases concentrations and trends are quoted for 1995, since concentrations change from year to year. For a few cases such data are unavailable as noted below.

1 a) For CH_3Cl and $CHCl_3$, data from a long-time series is used. The concentrations reported are averages of global measurements between 1985–1995. For the sites mentioned above the concentrations of CH_3Cl from Alaska to Antarctica are 572.7, 590.7, 608.5, 608, 591.6, 573.2 pptv, the lifetimes due to OH are 11.7, 2.2, .8, .9, 3.6, 12.3 years, and the estimated emissions are (0, 0.34, 1.68, 1.58, 0.13, 0.01) Tg/yr. For $CHCl_3$ the concentrations are 28.3, 32.8, 17.3, 9.7, 12.9, 17.6 pptv at the surface, and declines at 4–8%/Km to the tropopause, the lifetimes due to OH reactions are 3, 0.6, 0.2, 0.3, 1, 3.2 years, and the estimated emissions are (8, 282, 111, 6, 77, 13) Gg/yr (Khalil and Rasmussen 1997a, 1997c).

1 b) Hydrogen chloride (HCl) concentrations are taken from the analysis of Graedel and Keene (1995). No latitudinal differences could be quantified due to the high variability of the composite data; therefore, a simple average number is quoted for all the regions.

1 c) Dichloromethane (CH_2Cl_2) concentrations are from Koppmann et al. (1993) (digitized from graph) and phosgene ($COCl_2$) concentrations are from Singh et al. (1976; 1981) as averaged by Kindler et al. (1995).

1 d) Air and sea water concentrations of trichloroethylene (C_2HCl_3, TCE) and tetrachloroethylene (C_2Cl_4, PCE) were measured on the cruise of the Russian ship Academik Korolev, SAGA 1, during Fall 1983 covering latitudes from 45 N to 30 S. The flux estimates reported in the text are from this cruise. The concentrations of TCE were multiplied by an estimated calibration factor of 0.2 to put them on the scale of Koppmann et al. (1993). A similar factor of 0.2 is indicated for PCE to put these data on the scale reported by Wang et al. (1995). TCE data from the middle and high southern latitudes were not available, so the data from 0–30°S were used for this region. In most cases this is an accurate estimate, although it could be higher than actual concentrations. For PCE, the annual averaged data from Wang et al. (1995) are shown in the table obtained by digitizing the published graphs. The calibration difference is a sizable discrepancy between the measurements of different groups. The resolution of this discrepancy requires more research and could change the conclusions regarding the role of PCE and TCE in the global chlorine budget.

1 e) Dichloroethane ($C_2H_4Cl_2$) and bromochloromethane (CH_2BrCl) data are from Khalil and Rasmussen (1997b) based on the cruise of the Academik

Korolev, SAGA 2, during May–July 1987 covering latitudes from 50 °N to 42 °S.

1f) Methylchloroform (CH_3CCl_3) concentrations are the annual average values for 1995 taken from the six station network described above (see Khalil and Rasmussen 1984; Rasmussen and Khalil 1986). At middle and higher latitudes concentrations are lower in summer and fall and higher in the winters, driven by the seasonality of OH.

1g) F-22 results are from Montzka et al. (1993, 1996). The average concentrations are for 1995.

1h) Data on F-21 are from Penkett et al. (1980) and represent concentrations in 1979. It is not known whether there are trends that could make present concentrations substantially different.

1i) Data for F-12, F-11, F-10 (CCl_4), and F-113 are from the six sample network of Khalil and Rasmussen described earlier. Values reported are averages for 1995. These data are comparable with results from other networks such as the ALE/GAGE program and the NOAA/CMDL program (Simmonds et al. 1988; Prinn et al. 1992; Cunnold et al. 1994; Elkins et al. 1996; Montzka et al. 1996).

1j) F-114 and $CBrClF_2$ (Halon 1211) are from the SAGA 2 cruise discussed in note 1e. For Halon 1211 the values obtained on the cruise have been extrapolated by adding 0.15 pptv/yr (Khalil and Rasmussen 1986). The resulting concentrations are close to those in Montzka et al. 1996 (in the figures). The values reported, therefore, reflect 1995 levels.

1k) F-115 is from Fabian et al. (1996) at about 8 pptv for 1987. There is no information on latitudinal distribution hence the same value is reported for all latitudes. For F-13 no significant latitudinal difference was observed in the limited data available from Penkett et al. (1981). The average quoted is for 1979.

1l) The HCFC data are digitized from graphs in the papers by Elkins et al. (1995) and Montzka et al. (1996).

(2) **Trends.** A simple linear approach was adopted to estimate the trends as "b" in $C = Co + b\,t$, where C is the concentration in time t. Trends were estimated over the most recent times to show the current state of the chlorine budget. Past trends are discussed in the text. The ± values are the 90% confidence limits of the estimated trend (Snedecor and Cochran 1980). In most cases the data used for estimating the trends are described in the notes above for the concentrations (1a–1l).

2a) For the group of reactive gases, no significant trends can be documented for any of the gases. There are no sustained trends for CH_3Cl and $CHCl_3$ between 1985 and 1995 for which the most reliable global data are available (Khalil and Rasmussen 1997a, c). The variabilities in the concentrations of these gases are large enough that trends in the order of 1 pptv/yr for CH_3Cl and 0.5 pptv/yr for $CHCl_3$ could go undetected, but these values represent negligible changes for the budget of methyl chloride and small changes of chloroform. For HCl, Graedel and Keene (1995) have suggested that there may be an increase due to the interaction of man-made pollution and the natural production of HCl from sea salt. The magnitudes of such changes are not known at present. For phosgene, there are no documented trends for most of the precursors, so no trends can be calculated. It is also noteworthy that trends of reactive compounds are difficult to detect and quantify because of the variability of the measured concentrations, and that the trends are not sustained for long periods as they follow the changes in emissions, which are affected by many variables.

2b) For the gases in the intermediate reactivity group, clear trends are observed. Since $CHClF_2$ (F-22) is currently the chemical of choice for use in refrigeration and air conditioning, its use and emissions are increasing (AFEAS 1997). This fact is reflected in the atmospheric trends. The trends quoted here are from Montzka et al. (1993) for the period 1/1987–12/1992. Methylchloroform, on the other hand, is being phased out and its concentrations are falling rapidly as the industrial emissions fall. The trends quoted are from the data discussed in note 1g above for the period between 1/1994 and 12/1995. The other gases in this range of lifetimes are not expected to have trends that would be important to the reactive chlorine budget in the future.

2c) The main long-lived gases are the well-known chlorofluorocarbons F-11 and F-12 and related compounds. For most of these there are slow changes or downward trends at present. The trend estimates here are for the period between 1/1994 and 12/1995 based on the six site network described earlier (note 1j). The trends for F-114 and F-115 are based on a mass balance calculation (Sect. 2.1) and the estimates of emissions by AFEAS (1997) – there are not enough observational data to estimate these trends directly. Trends for Halon 1211 are from Montzka et al. (1996). The rate of increase is slowing down (Butler et al. 1992).

2d) Trends for the replacement compounds HCFC 141b and 142b are from data presented by Elkins et al. (1996). Although here the linear model may not be very accurate, the concentrations at present are small enough that using the linear model provides a sufficient measure of increases for most environmental applications.

(3) Average Chlorine. Most of the available observational data are at the earth's surface, especially for the reactive chlorine gases. The concentrations of these gases can decline significantly with altitude in the troposphere, especially if the lifetime is short and the emissions are the earth's surface. The average concentration in the troposphere is defined as:

$$C_{ave} = \frac{\int_0^{z_T} nC(z)dz}{\int_0^{z_T} n(z)dz} \tag{3.1}$$

where n(z) is the number density of air and C(z) is the mixing ratio. In the absence of observational data the following approximate functions are used which assume a constant destruction rate in the troposphere, emissions at the surface (no atmospheric formation) and a constant rate of vertical transport in the troposphere.

$$C(z) \approx C_o \exp \frac{1}{2H}\left[\sqrt{1+\frac{4H^2}{K\tau}}\right]z \tag{3.2}$$

$$n \approx n_o e^{-\frac{z}{H}} \tag{3.3}$$

where H is the scale height (about 7.5–8 km), K is the vertical transport coefficient (taken to be 10 m^2/s) and τ is the average tropospheric lifetime (see Table 1, column 6).

Observations show that this effect is small for gases with lifetimes greater than a few months and negligible for gases with lifetimes of a year or more as these gases become well mixed in the troposphere. The chlorine contributed to the troposphere by each gas is: C(Cl) = Cave (35.45/MW), where Cave is in pptv of Cl.

(4) Natural Chlorine Fraction. There are insufficient observational data to apportion the emissions of the reactive chlorine compounds between natural and anthropogenic sources. An approximate approach is adopted as discussed in the text and notes on lifetimes below. The natural concentrations are estimated as: C (natural) = S (natural)/S × Cave, where S is from Table 1, column 7, S(natural) is estimated from the southern hemisphere concentrations as discussed in the text, and Cave is from column 3.

(5) Stratospheric Average Chlorine. The amount of chlorine in the stratosphere is estimated by the following approximate equation:

$$C_s = \frac{C_T}{(1+\lambda H)} \tag{5.1}$$

where C_T (pptv) is the average concentration at the tropopause, H is the scale height in the stratosphere, and λ is the exponential rate of decrease of mixing ratio with height in the stratosphere (assuming the vertical profile can be so described). Values for λ estimated from the paper of Fabian et al. (1996) are (in 1/km): CH_3Cl (0.134), CH_3CCl_3 (0.146), F-22 (0.054), F-12 (0.104), F-11 (0.134), CCl_4 (0.146), F-113 (0.107), F-114 (0.0644), F-115 (0.034), F-13 (0.04), H1211 (0.14 estimated). For the reactive gases, λ was taken to be 0.15.

(6) Lifetimes. For the reactive group, the lifetime of $COCl_2$ is from Kindler et al. (1995) and for HCl from Graedel and Keene (1995). For $C_2H_4Cl_2$ the lifetime is a rough estimate based on the inter-hemispheric gradient assuming that there are no sources in the southern hemisphere. The lifetime is calculated as $\tau_T/(R-1)$ where τ_T is the inter-hemispheric transport time (~ 1 year) and R is the measured ratio of the northern hemisphere to southern hemisphere concentrations.

For the remaining gases the lifetimes are calculated based on the reactions with OH as:

$$\tau = \frac{\int_V nC\,dV}{\int_V nK_R[OH]C\,dV} \tag{6.1}$$

where n(z) is the number density of air, C is the concentration, K_R is the reaction rate constant of the gas with OH, [OH] is the number density of hydroxyl concentrations and V is the volume of integration (here the entire atmosphere). The lifetimes are calculated by approximations to evaluate these integrals. The OH fields are derived from the photochemical model described by Lu and Khalil (1991). The present version of the model provides two-dimensional OH fields for any time of the year and any latitude and altitude. The OH chemistry over oceans and land is further separated in the model. The average tropospheric concentration of OH is about 10^6 molecules/cm^3. For CH_3Cl and $CHCl_3$ a low resolution two-dimensional (latitude-altitude) model is used as described by Khalil and Rasmussen (1997a). Both tropospheric and stratospheric losses are included.

For the other gases, a simpler method is used that takes into account the latitudinal changes of OH in the four semi-hemispheres described in note 1 above. It also takes into account the decrease in temperature with height and the decrease in concentrations of the gases with height. This method is also used for the moderately reactive gases CH_3CCl_3 and F-22. For these calculations the stratospheric losses are not included. These are likely to be small for the reactive gases. The method was also applied to other HCFCs as these might be produced in the future. The results are $C_2HCl_3F_2$ (HCFC-122) = 0.8 years, $C_2HCl_2F_3$ (HCFC-123) = 1.2 years, HCFC-123a = 3.6 years, C_2HCl_4 (HCFC-124) = 5.4 years, $C_2H_2Cl_2F$ (HCFC-123b) = 3 years, $C_2H_2ClF_3$ (HCFC-133a) = 3.3 years, $C_3HF_5Cl_2$ (HCFC-225c) = 1.7 years, $C_3H_3F_2Cl$ (HCFC-243c) = 20 years, and $C_3HF_5Cl_2$ (HCFC-225c) = 5.4 years. The results quoted here are close to other published estimates (WMO 1995; IPCC,1995; Prather and Spivakovsky 1990; Nimitz and Skaggs 1990).

The lifetimes of the chlorofluorocarbons and Halon 1211 are from WMO (1995, chap 13) and IPCC (part I, chap. 5).

(7) **Emissions.** For reactive gases that are removed primarily by OH radicals, and do not have known trends, the amount removed per year is calculated and projected to be the amount that must be emitted from all sources. The amount removed is:

$$S = \int_V n\,K_R[OH]\,C\,dV \tag{7.1}$$

For CH_3Cl and $CHCl_3$, the same model as that used to calculate the lifetimes was used (see note 6 above). For C_2HCl_3, C_2Cl_4, CH_2Cl_2, $CHFCl_2$ (F-21), and CH_2BrCl, only the tropospheric losses are included. Stratospheric losses are expected to be small. The HCl source is taken from Graedel and Keene (1995) and the $COCl_2$ emissions are taken from Helas and Wilson (1992) and Kindler et al. (1995). The source of HCl is extremely uncertain as shown in the review by Gradel and

Keene (1995). For $C_2H_4Cl_2$ only rough calculations are possible based on the mass balance model (2-box) as: $S = C/\tau$, with τ as determined by the formula in Eq. 6.1.

Emissions of CH_3CCl_3 are from Midgley and McCullough (1995) projected to 1995 starting with their estimates for 1993. Emissions of CCl_4 are taken from Simmonds et al. (1988) and scaled to the emissions of F-11 and F-12 to estimate current emission rates (1995). Emissions of H-1211 are from McCulloch (1992). Emissions of the remaining gases in the "moderately reactive" and "long-lived" groups are taken from available industry estimates for 1995 (AFEAS 1997). All the emissions estimated here are for 1995, and represent a very major reduction in industrial emissions compared with earlier years (see text).

Natural emissions can be estimated as C(natural) (note 4)/C_{ave} (note 3)× emissions (note 7). These are rough estimates based on current information (see text for details).

(8) Ozone-Depletion Potentials and Global-Warming Potentials. Chlorine- and bromine-containing compounds are often classified according to their potential for depleting the ozone layer or causing global warming. These likelihoods are quantified by the ozone-depletion potentials (ODPs) and global-warming potentials (GWPs) relative to the effects of major contributors. For global warming the standard is carbon dioxide and for ozone depletion it is normally F-11. These potentials are based on the effect of a mole or kilogram of gas emitted relative to the effect of the same amount of the standard gas emitted. The GWPs are from WMO (1995, chap 13) and IPCC (1995, part I, chap 5). The ODPs for $COCl_2$, $CHCl_3$, C_2HCl_3, and C_2Cl_4 are from Kindler et al. (1995), for CH_2Cl_2 from Nimitz and Skaggs (1992), and for the rest of the gases from WMO (1995, chap 13), using the semi-empirical ODPs when available and model-derived values otherwise.

8
Discussion and Observations

Although there is very little information on many key reactive chlorine gases, some patterns have emerged from the data. An aspect of current interest is the amount and environmental characteristics of natural chlorine compared to the anthropogenic fraction. For the organic chlorine, with the data available, only rough estimates can be made to separate one from the other.

There is about 23 Tg of gaseous chlorine in the atmosphere making up nearly 4 ppbv of the troposphere. Of this only about 3 Tg (including HCl) can be identified with natural emissions, all of which are in the form of reactive chlorine gases. For the "reactive" gases category, methyl chloride has a demonstrated natural source and an observed global distribution that supports the importance of this source. The upper limit of the natural emissions is about 2.1 Tg Cl/yr assuming that the estimated global emissions are all natural except the biomass burning source of about 0.4 Tg Cl/yr. For other gases in this group there are data showing supersaturations in sea water for CH_2Cl_2, $CHCl_3$, C_2Cl_4, C_2HCl_3, and $C_2H_4Cl_2$. A certain amount of $COCl_2$ comes from the natural fractions of the precursors. Thus, most reactive gases have natural sources. Upper limits of the natural sources can be calculated assuming that all emissions in the southern hemisphere are natural, and that an equal amount of natural emissions would also exist in the northern hemisphere, as would be expected, for instance, from the oceanic source. The southern hemisphere source is estimated as $S_s = C_s/\tau + (C_n - C_s)/\tau_T$, (where τ_T is the average inter-hemispheric transport time, taken to be 1 year). For $COCl_2$, the calculated natural sources of the precursors are used according to the data from Kindler et al. (1995) on the yield of $COCl_2$ from each compound. The concentrations resulting from the natural emissions are summarized in Table 1 (Sect. 7).

The results show that for the "reactive" group, natural emissions of the organic chlorine gases are about 3.2 Tg Cl/yr or a little more than half of the total for this category. The natural sources constitute 75% of the chlorine concentration in the "reactive" group chlorine concentration (0.6 ppbv Cl or 3.1 Tg in the troposphere). This high value is dominated by CH_3Cl. If we compare these results with all chlorine, including the long-lived gases, the proportion of natural emissions does not change much because the long-lived gases are not emitted in large quantities, but the contribution to the atmospheric burden changes considerably down to about 15% as natural chlorine. A similar analysis using only CH_3Cl shows the contribution of natural chlorine to be about 0.3 Tg or about 13% of the total stratospheric organic chlorine from all sources.

These estimates suggest that natural organic chlorine is a sizable fraction of the total emissions of chlorine into the atmosphere, but that it is a relatively small fraction (~15%) of the standing concentration of chlorine in the troposphere and the stratosphere. We have not included the inorganic chlorine in the total emissions, otherwise the natural emissions would far exceed anthropogenic emissions, but still the standing concentration would be mostly from anthropogenic sources. It is the standing concentration in the troposphere that determines to a large extent whether the gases will cause global warming or ozone depletion.

It is interesting that all the natural chlorine emissions come from the "reactive" group of gases while gases with moderate or long atmospheric lifetimes are all anthropogenic. From the perspective of global warming and ozone depletion, the reactive chlorine gases are much safer than their long-lived counterparts produced by anthropogenic activities. The natural component, or indeed the whole reactive group, emitting large amounts of chlorine into the atmosphere, has a small, possibly negligible role in stratospheric ozone and climate. The replacement HCFCs behave similarly to the reactive gases discussed here and are thus expected to reduce ozone depletion and global warming.

Acknowledgements. I have benefitted from discussions with many colleagues, most notably Professor Rei Rasmussen (OGI), Martha J. Shearer (Portland State University), W.C. Keene (University of Virginia), T.M. Graedel (Yale University), David Erickson III (National Center for Atmospheric Research), R.M. Moore (Dalhousie University), Jürgen Lobert (Scripps Institute of Oceanography), M.L. Aucott, and Archie McCullough (ICI). This work was supported in part by the Chemical Manufacturers Association (Chlorine Chemistry Council) and European Chemical Industry Council (Euro Chlor). Some of the research used in this study was supported by the National Science Foundation (NSF DPP-8821320, ATM 88-11059), Department of Energy (DE-FG06-84ER60613). Additional support was provided by Andarz Co.

9
References

Alternative Fluorocarbon Environmental Acceptability Study (AFEAS), Production, Sales and Atmospheric Release of Fluorocarbons Through 1995 (AFEAS, Science and Policy Services, Inc, West Tower - Suite 400, 1333 H Street N.W., Washington DC, 20005), 1997

Andrea MO (1991) Biomass burning: its history, use and distribution and its impact on environmental quality and global climate In: Levine JS (ed) Global Biomass Burning. MIT Press, Cambridge, MA

Atlas E, Pollock W, Greenberg J, Heidt L, Thompson AM (1993) Alkyl nitrates, nonmethane hydrocarbons, and halocarbon gases over the equatorial Pacific Ocean during SAGA 3. J Geophys Res 98:16933

Aucott M, Graedel TE, Klienman G, McCulloch A, Midgley P, Li Y-F (1997) Anthropogenic emissions of chloroform and chloromethane: Reactive chlorine emissions inventory. Submitted for publication

Brasseur G, Hitchman MH, Walters S, Dymek M, Falise E, Pirre M (1990) An interactive chemical dynamical radiative two-dimensional model of the middle atmosphere. J Geophys Res 95:5639

Butler JH, Elkins JW, Hall BD, Cummings SO, Montzka SA (1992) A decrease in the growth rate of atmospheric halon concentrations. Nature 358:403

Cadle RA (1980) A comparison of volcanic with other fluxes of atmospheric trace constituents. Rev Geophys Space Phys 18:746

Ciccioli P et al. (1976) Evaluation of organic pollutants in the open air and atmospheres in industrialized sites using graphatized carbon black traps and gas chromatographic-mass spectrometric analysis with specific detectors. J Chromatogr 126:757

Class T, Ballschmiter K (1987) Atmospheric halocarbons: global budget estimations of tetrachloride, 1,2-dichloroethane, 1,1,1,2-tetrachloroethane, hexachloroethane and hexachlorobutadiene. Estimation of hydroxyl radical concentrations in the troposphere of the northern and southern hemispheres Fresenius Z Anal Chem 327:198

Class T, Ballschmiter K (1986) Chemistry of organic traces in air VI: Distribution of chlorinated C_1-C_4 hydrocarbons in air over the northern and southern Atlantic ocean. Chemosphere 15:413

Cowan MI et al. (1973) Production of volatile metabolites by species of foams. Trans Br Mycol Soc 60:347

Crutzen PJ, Heidt LE, Krasnec JP, Pollock WH, Seiler W (1978) Biomass burning as a source of atmospheric trace gases CO, H_2, N_2O, NO, CH_3Cl, and COS. Nature 282:253

Cunnold DM et al. (1994) Global trends and annual releases of CCl_3F and CCl_2F_2 estimated from the ALE/GAGE and other measurements from July 1978 to June 1991. J Geophys Res 99:1107

Denovaan BA, Strand SE (1992) Biological degradation of chlorofluorocarbons in anaerobic environments. Chemosphere 24:935

Edgerton SA, Khalil MAK, Rasmussen RA (1986) Source emission characterization of residential wood burning stoves and fireplaces: Fine particle to methyl chloride ratios for use in chemical mass balance modeling. Environ Sci Technol 20:803

Elkins JW (ed) (1996) Nitrous oxide and halocarbons, Climate Monitoring and Diagnostics Laboratory. Report No. 23 (NOAA/CMDL, 325 Broadway, Boulder, 80303

Fabian P, Borchers R, Penkett SA, Prosser NJD (1981) Halocarbons in the stratosphere Nature 294:733

Fabian P, Borchers R, Leifer R, Subbaraya BH, Lal S, Boy M (1996) Global stratospheric distribution of halocarbons. Atmos Environ 30:1787

Fabian P, Borchers R, Gomer D, Penkett SA (1984) Vertical profiles of halocarbons in the stratosphere. Adv Space Res 4:135

Fabian P, Gomer D (1984) The vertical distribution of halocarbons in the stratosphere. Fresenius Z Anal Chem 319:890

Fabian P et al. (1981) The vertical distribution of stable trace gases at mid latitudes. J Geophys Res 86:5179

Fabian P, Borchers R, Kruger BC, Lal S (1985) The vertical distribution of $CHClF_2$ (CFC-22) in the stratosphere. Geophys Res Lett 12:1
Fabian P, Borchers R, Kruger BC, Lal S (1985) The vertical distribution of CFC-114 in the atmosphere. J Geophys Res 90:13091
Froidevaux LM, Allen, Yung YL (1985) A critical analysis of ClO and O_3 in the mid-latitude stratosphere. J Geophys Res 90:12999
Galloway JN, Likens GE, Keene WC, Miller JM (1982) The composition of precipitation in remote areas of the world. J Geophys Res 87:8771
Goldan PD, Kuster WC, Albritton DL, Schmeltokopf AL (1980) Stratospheric $CFCl_3$, CF_2Cl_2 and N_2O height profile measurements at several latitudes. J Geophys Res 85:413
Graedel TE (1978) Chemical compounds in the atmosphere. Academic Press, New York
Graedel TE, McGill R (1986) Degradation of materials in the atmosphere. Environ Sci Technol 20:1093
Graedel TE, Hawkins DT, Claxton LD (1986) Atmospheric chemical compounds. Academic Press, New York
Graedel TE, Keene WC (1995) Tropospheric budget of reactive chlorine, Global Biogeochemical Cycles, 9:44
Harlov R, Kebbekus B, Bozzelli JW, Lioy PJ (1983) Measurement of selected volatile organic compounds at three locations in New Jersey during the summer season. JAPCA 33:1177
Harper DB (1985) Halomethane from halide ion – a highly efficient fungal conversion of environmental significance. Nature 315:55
Helas G, Wilson SR (1992) On sources and sinks of phosgene in the troposphere. Atmos Environ 26A, 2875
Hummel JR, Reck RA (1981) The direct thermal effects of $CHClF_2$, CH_3CCl_3 and CH_2Cl_2 on the surface temperatures. Atmos Environ 15:379
Intergovernmental Panel on Climate Change (IPCC), Climate Change 1994: Radiative Forcing of Climate Change. Cambridge University Press, Cambridge
Jobson BT, Niki H, Yokouchi Y, Bottenheim J, Hopper F Leaith R (1994) Measurement of C_2–C_6 hydrocarbons during the polar sunrise 1992 experiment: evidence for Cl atom and Br atom chemistry. J Geophys Res 99:25355
Keene WC et al. (1990) The geochemical cycling of reactive chlorine through the marine troposphere. Global Biogeochem Cycles 4:407
Keene WC, Jacob DJ, Fan S-M (1996) Reactive chlorine: a potential sink for dimethyl sulfide and hydrocarbons in the marine boundary layer. Atmos Environ 30:i–iii
Khalil MAK, Rasmussen RA (1984) Modeling chemical transport and mass balances in the atmosphere. In: Neely WB, Blau GE (eds) Environmenal exposure from chemicals, vol II. CRC Press, Boca Raton, Florida, pp 21–54
Khalil MAK, Rasmussen RA (1997a) Atmospheric methyl chloride, (Report No. 01–997, Department of Physics, Portland State University, Portland, OR.) Atmospheric Environment (1999 – in the press)
Khalil MAK, Edgerton SA, Rasmussen RA (1983) A gaseous tracer model for air pollution from residential wood burning. Environ Sci Technol 17:555
Khalil MAK, Rasmussen RA (1997c) Atmospheric Chloroform (Report No. 02–997, Department of Physics, Portland State University, Portland, OR.) Atmospheric Environment (1999 – in the press)
Khalil MAK, Rasmussen RA (1997b) Ocean-air exchange of atmospheric trace gases (Report No. 01–1097, Department of Physics, Portland State University, Portland, OR.) Submitted for publication
Khalil MAK, Rasmussen RA (1986) The trend of bromochlorodifluoromethane and the concentrations of other bromine-containing gases at the South Pole Antarctic J United States 20:206
Khalil MAK, Rasmussen RA, Hoyt SA (1983) Atmospheric chloroform: ocean-air exchange and global mass balance. Tellus 35B:266
Khalil MAK, Rasmussen RA (1993) Arctic haze: patterns and relationships to regional signatures of trace gases. Global Biogeochemical Cycles 7:27

Khalil MAK, Rasmussen RA (1993) The environmental history and probable future of fluorocarbon 11. J Geophys Res 98:23091
Khalil MAK, Rasmussen RA (1989) The potential of soils as a sink of chlorofluorocarbons and other man-made chlorocarbons. Geophys Res Lett 16:679
Khalil MAK, Rasmussen RA (1984) The atmospheric lifetime of methylchloroform (CH_3CCl_3) Tellus 36B:317
Kindler TP, Chameides WL, Wine PH, Cunnold DM, Alyea FN, Franklin JA (1995) The fate of atmospheric phosgene and the stratospheric chlorine loadings of its parent compounds: CCl_4, C_2Cl_4, C_2HCl_3, CH_3CCl_3, and $CHCl_3$. J Geophys Res 100:1235
Koppmann R, Johnen FJ, Plass-Dulmer C, Rudolph J (1993) Distribution of methyl chloride, dichloromethane, trichloroethene and tetrachlorethene over the north and south Atlantic. J Geophys Res 98:20517
Lal S, Borchers R, Fabian P, Kruger BC (1985) Increasing abundance of $CBrClF_2$ in the atmosphere. Nature 316:135
Laux JM, Hemminger JC, Finlayson-Pitts BJ (1994) X-ray photoelectron spectroscopic studies of the heterogeneous reaction of gaseous nitric acid with sodium chloride: Kinetics and contribution to the chemistry of the marine troposphere. Geophys Res Lett 21:1623
Lillian D et al. (1975) Atmospheric fates of halogenated compounds. Environ Sci Technol 12:1042
Lovelock JE (1977) Methylchloroform in the troposphere as an indicator of OH radical abundance. Nature 267:32
Lu Y, Khalil MAK, Tropospheric OH (1991) Model calculations of spatial, temporal and secular variations. Chemosphere 23:397
McCulloch A (1992) Global production and emissions of bromochlorodifluoromethane and bromotrifluoromethane (Halons 1211 and 1301). Atmos Environ 26A:1325
Midgley P, McCulloch A (1995) The production and global distribution of emissions to the atmosphere of 1,1,1-trichloroethane (methylchloroform). Atmos Environ 29:1601
Molina MJ, Rowland FS (1974) Stratospheric sink for chlorofluoromethanes: chlorine atom catalyzed destruction of ozone. Nature 249:810
Montzka SA et al. (1996) Decline in the tropospheric abundance of halogen from halocarbons: implications for stratospheric ozone depletion. Science 272:1318
Montzka SA, Myers RC, Butler JH, Elkins JW (1993) Global tropospheric distribution and calibration scale of HCFC-22. Geophys Res Lett 20:703
Moore RM, Groszko W, Niven SJ (1996) Ocean-atmosphere exchange of methyl chloride: results from NW Atlantic and Pacific Ocean studies. J Geophys Res 101:28529
NAS, Chloroform, Carbon Tetrachloride, and Other Halocarbons: An Environmental Assessment. National Academy of Sciences, Washington DC, 1078
NASA (1997) Chemical Kinetics and Photochemical Data for Use in Stratospheric Modeling, Evaluation No.12, JPL, Pasadena, CA
National Safety Council, Chemical Backgrounders - Environment Writer (NSC, Environmental Health Center, 1025 Connecticut Av., NW, Suite 1200, Washington DC, 20036, 1997)
Nimitz JS, Skaggs RS (1992) Estimating tropospheric lifetimes and ozone depletion potentials of one and two carbon hydrofluorocarbons and hydrochlorofluorocarbons. Environ Sci Technol 26:739
Pellizzari ED (1982) Analysis of organic vapor emissions near industrial and chemical waste disposal sites. Environ Sci Technol 16:781
Pellizzri ED, Bunch JE, Berkley RE, McRae J (1976) Determination of trace hazardous organic vapor pollutants in ambient atmospheres by gas chromatography/mass spectrometry/computer. Anal Chem 48:803
Penkett SA, Prosser NJD, Rasmussen RA, Khalil MAK (1981) Atmospheric measurements of CF_4 and other fluorocarbons containing the CF_3 group. J Geophys Res 86
Prather M, Spivakovsky CM (1990) Tropospheric OH and the lifetimes of hydroclhorofluorocarbons. J Geophys Res 95:18723

Prinn R et al. (1992) Global average concentration and trend for hydroxyl radicals deduced from ALE/GAGE trichloroethane (methylchloroform) data from 1978–1990. J Geophys Res 97:2445
Prinn et al. (1995) Atmospheric trends and lifetime of CH_3CCl_3 and global OH concentrations. Science 269:187
Pszenny A et al. (1993) Evidence of inorganic chlorine gases other than hydrogen chloride in marine surface air. Geophys Res Lett 20:699
Rasmussen RA, Rasmussen LE, Khalil MAK, Dalluge RW (1980) Concentration distribution of methyl chloride in the atmosphere. J Geophys Res 85:7350
Rasmussen RA, Khalil MAK (1982) Latitudinal distributions of trace gases in and above the boundary layer. Chemosphere 3:227
Rasmussen RA, Khalil MAK (1986) Atmospheric trace gases: trends over the last decade. Science 232 1623
Rice DW, Cappell RJ, Kinsolving W, Laskowski JJ (1980) Indoor concentrations of metals. J Electrochem Soc 127:891
Rudolph J, Khedim A, Koppmann R, Bonsang B (1995) Field study of the emissions of methyl chloride and other halocarbons from biomass burning in Western Africa. J Atmos Chem 22:67
RudolphJ, Koppermann R, Plass-Dulmer C (1996) The budgets of ethane and tetrachloroethene: is there evidence for the impact of reactions with chlorine atoms in the troposphere. Atmos Environ 30:1887
Sander R, Crutzen PJ (1996) Model study indicating halogen activation and ozone destruction in polluted air masses transported to sea. J Geophys Res 101:9121
Simmonds PG et al. (1988) Carbon tetrachloride lifetimes and emissions determined from daily global measurements during 1978–1985. J Atmos Chem 7:35
Singh HB, Thakur AN, Chen YE, Kanakidou M (1996a) Tetrechloroethylene as an indicator of low Cl atom concentrations in the troposphere. Geophys Res Lett 23:1529
Singh HB et al. (1996b) Low ozone in the marine boundary layer of the tropical Pacific ocean: Photochemical loss, chlorine atoms and entrainment. J Geophys Res 101:1907
Singh HB, Salas LJ, Stiles RE (1983b) Methyl halides in and over the Eastern Pacific (40oN-32oS). J Geophys Res 88:3684
Singh HB (1976) Phosgene in ambient air. Nature 264:428
Singh HB, Salas LJ, Shigeishi H, Scribner E (1979) Atmospheric halocarbons, hydrocarbons and sulfur hexafluoride: Global distributions, sources and sinks. Science 203:899
Singh HB, Salas LJ, Stiles RE (1982) Distribution of selected gaseous organic mutagens and suspect carcinogens in ambient air. Environ Sci Technol 16:872
Singh HB, Salas LJ, Smith AJ, Shigeishi H (1981) Measurements of some potentially hazardous organic chemicals in urban environments. Atm Environ 15:601
Singh HB, Salas LJ, Stiles RE (1983a) Selected man-made halogenated chemicals in the air and oceanic environment. J Geophys Res 88:3675
Singh HB (1977a) Preliminary estimation of average tropospheric HO concentrations in the northern and southern hemispheres. Geophys Res Lett 453
Singh HB, Salas LJ, Cavanagh LA (1977c) Distribution, sources and sinks of atmospheric halogenated compounds. JAPCA 27:332
Singh HB, Salas LJ, Shigeishi H, Crawford A (1977b) Urban-nonurban relationships of halocarbons, SF6, N_2O, and other trace constituents. Atmos Environ 11:819
Snedecor GW, Cochran WG (1980) Statistical methods. Iowa State University Press, Ames, Iowa
Tait VK, Moore RM, Tokarczyk R (1994) Measurements of methyl chloride in the northwest Atlantic. J Geophys Res 99:7821
Vierkorn-Rudolph B, Bachmann K, Schwarz B, Miexner FX (1984) Vertical profiles of hydrogen chloride in the troposphere. J Atmos Chem 2:47
Vogt R, Crutzen PJ, Sander R (1996) A mechanism for halogen release from sea salt aerosol in the remote marine boundary layer. Nature 383:327
Wang CJ-L, Blake DR, Rowland FS (1995) Seasonal variations in the distribution of a reactive chlorine compound, tetrachloroethene ($CCl_2 = CCl_2$). Geophys Res Lett 22:1097

Warneck P (1988) The chemistry of the natural atmosphere. Academic Press, New York
White RH (1982) Analysis of dimethylsulfonium compounds in marine algae. J Mar Res 40:529
Wiedmann TO, Guthner B, Class TJ, Ballschmiter K (1994) Global distribution of tetrachloroethene in the troposphere: measurements and modeling. Environ Sci Technol 28:2321
Wingetner OW, Kubo MK, Blake NJ, Smith TW Jr, Blake DR, Rowland FS (1996) Hydrocarbon and halocarbon measurements as photochemical and dynamical indicators of atmospheric hydroxyl, atomic chlorine and vertical mixing obtained during Lagrangian flights. J Geophys Res 101:4331
World Meteorological Organization (WMO) (1985) Atmospheric Ozone. Global Ozone Research and Monitoring Project, Report No. 16, Geneva, Switzerland
World Meteorological Organization (WMO) Scientific Assessment of Ozone Depletion (Global Ozone Research and Monitoring Project, Reports No. 20, 25, 37, Geneva, Switzerland, 1989, 1991, 1995)
World Meteorological Organization (WMO) (1988) Report of the International Ozone Trends Panel Global Ozone Research and Monitoring Project, Report No.18, Geneva, Switzerland
Yokouchi Y, Barrie LA, Toom D, Akimoto, H (1996) The seasonal variation of selected natural and anthropogenic halocarbons in the arctic troposphere. Atmos Environ 30:1723

CHAPTER 3

Reactive Fluorine Compounds

J. Harnisch

Max-Planck-Institut für Aeronomie, Max-Plank-Str. 2, 37191 Katlenburg-Lindau, Germany.
E-mail: harnisch@linmpi.mpg.de

Human activities have significanty increased the atmospheric concentrations of various compounds of fluorine. The input of such compounds into the environment is expected to grow rapidly in the future. Their environmental fate and impact should thus be carefully assessed. Compounds of fluorine are covered by this chapter if they do not contain other halogens and exhibit at least some economic or environmental relevance. Here, these substances are divided into the following groups: hydrofluorocarbons, perfluorocarbons, further organic fluorine compounds, long- and short-lived inorganic fluorine compounds. Their respective environmental relevance is discussed with special attention to the issues of "global warming", "ozone depletion" and to their and their degradation products' toxic effects on the biosphere. A revised picture of the natural geochemistry of atmospheric fluorine is sketched including volcanic emissons, mineral dust, biomass burning, the release from sea-salt aerosol and the degassing of natural tetrafluoromethane. This natural situation is then compared with the anthropogenically disturbed case and a projected scenario for the year 2050.

Keywords: Fluorine, atmosphere, geochemistry, accumulation, degradation.

The Handbook of Environmental Chemistry Vol. 4 Part E
Reactive Halogen Compounds in the Atmosphere
(ed. by P. Fabian and O. N. Singh)

1 Introduction

During the last decades atmospheric fluorine-compounds have caused considerable environmental concern. First, industrial emissions of hydrogen fluoride and other inorganic fluoride compounds with their local toxic effects on plants, animals and humans initiated systematic research on the distribution and fate of atmospheric fluorine. Secondly, it was the issue of the depletion of polar stratospheric ozone that called for research on a possible contribution of degradation products of chlorofluorocarbons (CFC) to the catalytic destruction of ozone. Thirdly, as worries grew about the effects of a possible anthropogenic warming of the global climate system, the use and release of radiative fluorine compounds started to be discussed controversially. Fourthly, concern was recently expressed about the possible environmental accumulation of trifluoroacetic acid (TFA), which is a toxic degradation product of some of those hydrofluorocarbons intended to rapidly replace many ozone depleting CFC in their traditional applications.

This chapter exclusively deals with those atmospheric compounds that do not contain any other halogen apart from fluorine. For the major mixed halogen-compounds readers should therefore refer to the other chapters of this volume.

A large group of organic fluorine compounds is produced only in small quantities and could not be covered in this chapter. The full range of these fluorinated compounds that are industrially produced today is described in books on industrial chemistry like Ullmann [1, 2, 3]. The selection of substances for this chapter must remain somewhat arbitrary as it was based on the availability of data and measurements. There may well be several relevant fluorine-compounds being released naturally or industrially which have yet been overlooked by those involved with the monitoring of the Earth's atmosphere.

The major aim of this contribution was to compile the available information upon early 1997 on the production, release, and the atmospheric distribution and fate of the substances selected. Additionally it was tried to resume earlier attempts to gain a more quantitative understanding of the atmospheric geochemistry of fluorine including a large set of natural and anthropogenic substances.

2 Atmospheric Fluorine Compounds

Man has been using fluorine and its minerals for many centuries [4]. The knowledge about its main compounds had been established during the nineteenth century. It took until the middle of this century for the commercial breakthrough of organic fluorine chemistry. During the 1990s hydrofluorocarbons (HFC) have started to replace other halocarbons that had to be phased out because of their adverse effects on stratospheric ozone. This trend will probably continue. Unlike the other halogens chlorine, bromine and iodine it is fluorine that will show a rapidly increasing atmospheric abundance well into the twenty-first century.

2.1 Hydrofluorocarbons

Until the late 1980s (hydro)chlorofluorocarbons ((H)CFCs) had reached a remarkable commercial success in many applications (Fig. 1). Despite their similar thermodynamic properties as well as being odorless and non-toxic the class of HFCs had only very limited technical relevance due to their higher price. Flammability also made some HFCs appear to be an inferior technical solution compared to (H)CFCs. The environmental properties of HFCs are on the other hand superior to (H)CFCs as they lack chlorine and in general exhibit shorter atmospheric lifetimes (Table 1) [5–7]. Also, HFCs do not contribute significantly to the photochemical formation of tropospheric ozone [8]. The international agreement of the Montreal Protocol (1987) [9, 10] and the following Accords of London (1990), Copenhagen (1992) and Vienna (1995) [10] restricted the use of most chlorinated halocarbons in order to protect the stratospheric ozone layer. Just like the (H)CFCs also HFCs are potent greenhouse gases (Table 1) [22]. In most fields of applications such as solvents, aerosol propellants, foam blowing and refrigeration alternative technologies are now readily available to permanently replace halocarbons. Nevertheless, in many cases it was preferred to

Table 1. Properties of the some of the most common fluorinated compounds. Given are the chemical formula of the substance, its trade code, molar mass, boiling point, atmospheric lifetime, the absolute radiative forcing at the tropopause for average cloudiness and the global warming potential for a 100 year integration period relative to CO_2 on a mass-basis

Formula	code	mol. mass [g mol^{-1}]	boiling point [°C]	atmosph. lifetime [yr]	rad. forcing [W m^{-2} ppb^{-1}]	GWP-100 rel. CO_2
CH_3F	HFC41	34	-78.5 [3]	3.7 [126]	0.03 [22]	150 [126]
CH_2F_2	HFC32	52	-51.7 [3]	5.6 [126]	0.14 [22]	650 [126]
CHF_3	HFC23	70	-82.1 [3]	264 [126]	0.23 [22]	11,700 [126]
CF_4	FC14	88	-128 [3]	> 50,000 [127]	0.10 [129]	6,500 [126]
CH_3CH_2F	HFC161	48	-37.1 [3]	0.3 [7]		
CH_3CHF_2	HFC152a	66	-24.7 [3]	1.5 [126]	0.13 [22]	140 [126]
CH_3CF_3	HFC143a	84	-47.6 [3]	48.3[126]	0.17 [22]	3,800 [126]
CH_2FCF_3	HFC134a	102	-26.3 [3]	14.6 [126]	0.18 [22]	1,300 [126]
CHF_2CF_3	HFC125	120	-48.5 [3]	32.6 [126]	0.25 [22]	2,800 [126]
C_2F_6	FC116	138	-78.1 [3]	> 10,000 [127]	0.22 [129]	9,200 [126]
$CF_3CH_2CF_3$	HFC236fa	152		209 [126]	0.27 [22]	6,300 [126]
CF_3CHFCF_3	HFC227a	170	-16.5 [128]	36.5 [126]	0.28 [22]	2,900 [126]
C_3F_8	FC218	188	-36.7 [3]	2,600 [126]	0.23 [129]	7,000 [126]
c-C_4F_8	RC318	200	-6.1 [3]	3,200 [126]	0.32 [130]	8,700 [126]
SF_6		146	-63.9 [131]	3,200 [126]	0.64 [130]	23,900 [126]
NF_3		71	-129 [90]	> 500 [91]		8,000 [91]

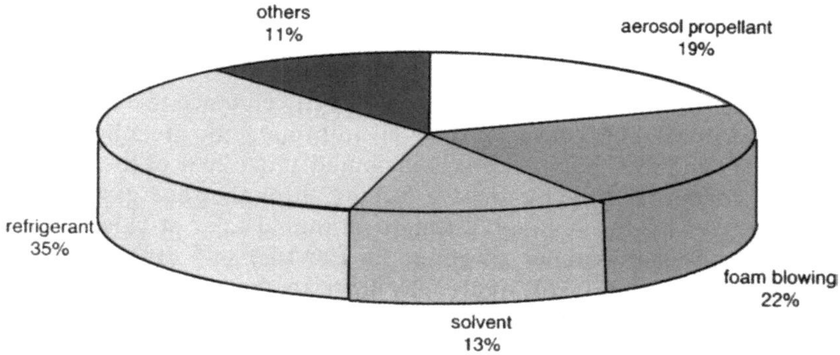

Fig. 1. Pre-Montreal era – estimate of worldwide uses of (hydro) chlorofluorochlorocarbons 11, 12, 113, 114 and 22. The world market was 1,3000,000 t yr^{-1} in 1986 [136, 137]; applications in 1990 after [14]

use HFCs as direct replacements of (H)CFCs for reasons of profitability, safety or convenience. The world market for fluoro-organic compounds will remain significantly smaller (e.g. 40% of historic CFC-usage [10]) in the foreseeable future than it was for (H)CFCs during the late 1980s before international regulations came into effect. It will depend greatly on the decisions of consumers, engineers and governments whether the future consumption of HFCs will be as high as predicted by the chemical industry [25]. Figure 2 shows the projected

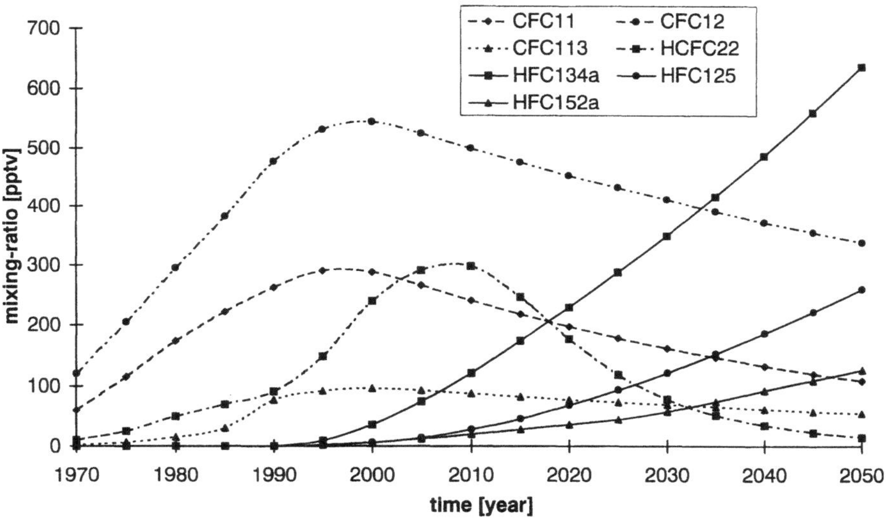

Fig. 2. Projected slow decrease of atmospheric abundance of CFCs 11, 12, 113 and of HCFC 22 and predicted increase of replacement HFCs 134a, 152a and 125 according to IPCC scenario IS92de [126]

tropospheric abundance of major halocarbons if HFC-emissions remain unregulated. While values for (H)CFCs slowly start to decrease after their emissions have been reduced, substances of the HFC-class could rapidly accumulate in the atmosphere. This could significantly enhance the absolute and relative contribution of halocarbons to the anthropogenic greenhouse effect. Between 1980 and 1990 halocarbons contributed above 20% of the total additional anthropogenic direct radiative forcing through trace gas emissions (0.054 Wm^{-2} yr^{-1}) [126]. A rough estimate of annual sales of halocarbons in 1986 yields a global number of about 2×10^{10} US$ yr^{-1} (10–20 $/kg [14]; 1,300,000 t yr^{-1} [136, 137]. This figure may illustrate the economic relevance of halocarbons. Global fossil fuel sales in terms of world-market prices in comparison were in the order of 2×10^{12} US$ yr^{-1} in 1986 being responsible for about 55% of the direct additional anthropogenic radiative forcing for the period 1980–1990. A simple analysis in terms of additional radiative forcing per US$ of sales reveals a greenhouse contribution in respect to economic relevance for halocarbons (5×10^{-13} W m^{-2} US$$^{-1}$) which is higher by a factor 25 than the corresponding value for fossil fuels (2×10^{-14} W m^{-2} US$$^{-1}$). An explicit economic cost-benefit analysis could of course yield differing results.

2.1.1
Monofluoromethane (CH_3F/HFC41)

Monofluoromethane currently has only very limited economic relevance. It is used in plasma etching processes of semiconductor production [11]. There are currently no atmospheric measurements available. All three other methylhalides are natural constituents of the Earth's atmosphere (today: 600 pptv CH_3Cl, 12 pptv CH_3Br, 1 pptv CH_3I). It is therefore likely that monofluoromethane can also be found in the atmosphere with a mixing-ratio of some parts per trillion. A possible source could be the terrestrial vegetation and its combustion products. Methane-oxidizing bacteria can on the other hand degrade CH_3F [12].

2.1.2
Difluoromethane (CH_2F_2/HFC32)

Difluoromethane (CH_2F_2) has recently gained commercial relevance. It is used in blends of different HFCs containing up to 50% HFC32 as a replacement of difluorochloromethane ($CHClF_2$/HCFC22) in stationary cooling equipment [13]. Two such proposed blends containing HFC32 are named "407C" and "410 A". Their composition together with their radiative properties is given in Table 2. It is currently unclear what share these blends will conquer of the total market for stationary refrigeration. McCulloch predicts a worldwide annual production of HFC32 of about 100,000 t yr^{-1} in 2020 [25]. Currently, atmospheric observations of HFC32 are unavailable.

Table 2. Composition and global warming potentials of blends 407C, 410A, 404 and 507 for stationary cooling [13]

	407C	410A	404A	507
HFC134a	52%	0%	4%	0%
HFC125	25%	50%	44%	50%
HFC143a	0%	0%	52%	50%
HFC32	23%	50%	0%	0%
Global Warming Potential rel. CO_2 for 100 years	1,525	1,725	3,260	3,300

2.1.3 *Trifluoromethane* (CHF_3/HFC23)

This substance has obtained some limited commercial relevance in applications for stationary very low temperature refrigeration [13, 14]. HFC23 is the most stable of all HFCs and exhibits an atmospheric lifetime of about 400 years [5, 6, 7]. Stratospheric profiles of HFC23 have been reported [49]. At tropopause-level in 1993 about 4–5 pptv HFC23 [49] were found corresponding to an atmospheric reservoir of about 50,000 t. CHF_3 may possibly be formed as a microbial degradation product of fluoroacetic acids in sediments [61].

2.1.4 *Monofluoroethane* (CH_3CH_2F/HFC161)

Monofluoroethane has practically no economic relevance. There are currently no atmospheric measurements available.

2.1.5 *Difluoroethane* (CH_3CHF_2/HFC152a)

HFC152a is currently only of limited economic importance. This may change to some degree as it is under consideration for use in stationary cooling and as blowing agent in closed cell foams [14]. Atmospheric observations of difluoroethane are currently unavailable.

2.1.6 *Trifluoroethane* (CF_3CH_3/HFC143a)

This substance is in the progress of reaching some commercial relevance. It is part of a binary (507) and a ternary (404A) blend for stationary refrigeration (see Table 2). Both blends have proven to successfully replace HCFC22 ($CHClF_2$) as well as the old blend 502, which was a mixture of 48.8% HCFC22 and 51.2% CFC115 (C_2ClF_5) [15]. In the future it may also so be used as a blowing agent for closed cell foams substituting the chlorinated HFCs still in use [14]. Atmospheric observations of HFC143a so far have not been reported.

2.1.7
Tetrafluoroethane (CH_2FCF_3/HFC134a)

Definitely HFC134a is the environmentally as well as economically most important HFC at present. It is replacing the banned CFCs in many applications. Its main applications will be as a refrigerant in mobile and stationary cooling, where it will probably dominate the market [13, 14] until halocarbon-free refrigeration technologies [16] are accepted more widely. HFC134a is also used as a blowing agent for the production of closed as well as open cell foams [14]. In open applications such as aerosol propellants hydrocarbons are generally accepted as preferable solutions. The use of HFC134a will here thus be less significant than previously estimated [14]. The environmental fate of HFC134a was studied extensively [17–19]. HFC134a is apparently not depleting stratospheric ozone [20, 21]. As in the case of the other HFCs it would progressively contribute to global warming [22] as its atmospheric concentration rises. The potential accumulation of the major atmospheric degradation product of HFC134a, TFA (see Sect. 2.3.1), in wetlands [60] is another issue of concern. Atmospheric observations of HFC134a were unavailable prior to 1996, when Montzka et al. [23] and Oram et al. [24] presented their measurements of the global distribution of HFC134a. Both groups independently report an exponential increase of 200% yr^{-1} of the tropospheric abundance of HFC134a between 1992 and mid-

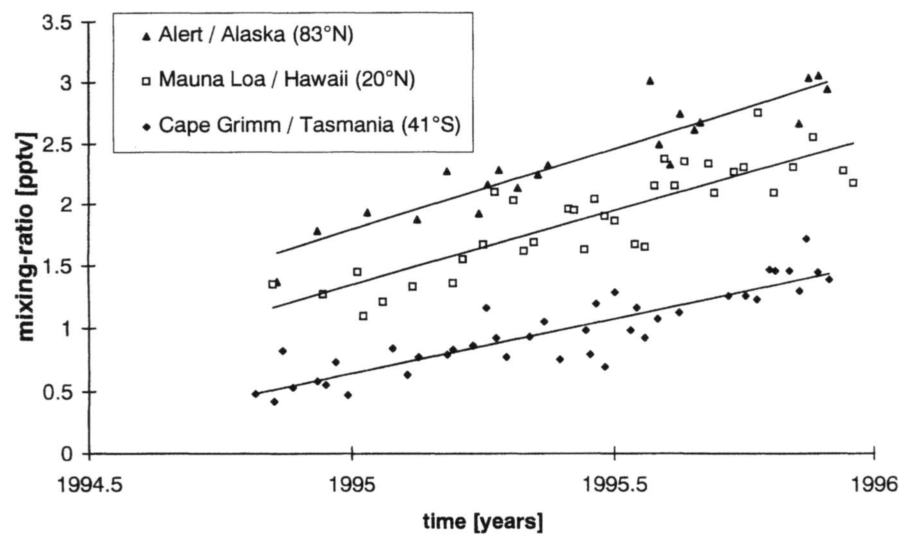

Fig. 3. Measured increase of the tropospheric abundance of HFC134a in 1994/95 for different locations. Data from NOAA/CMDL/after [23]

Table 3. Reported global sales and observed emissions of HFC134a

Year	Accountant's estimate of annual sales [132]	Observational estimate of annual emissions [24]
1990	189 t yr^{-1}	100 t yr^{-1}
1991	2,198 t yr^{-1}	250 t yr^{-1}
1992	6,404 t yr^{-1}	800 t yr^{-1}
1993	26,526 t yr^{-1}	2,900 t yr^{-1}
1994	50,400 t yr^{-1}	8,300 t yr^{-1}
TOTAL	85,717 t	12,500 t

1995. In the northern hemisphere (53 °N) an average volume mixing ratio of 2.2 pptv was observed for mid-1995, while only 0.8 pptv were found in the southern hemisphere (40 °S). This strong interhemispheric gradient is due to the unequal global distribution of emissions and the slow exchange of air-masses between both hemispheres (Fig. 3). Using a two-dimensional model of atmospheric transport Oram et al. [24] could estimate annual global emissions of HFC134a, which can be compared with reported sales [132] (Table 3). These data currently exhibit a two year delay, which is likely to increase in the future as gains of the world production of HFC134a slow down and emissions from long-lived applications become more relevant. McCulloch [25] predicts a quasi-linear increase of the global production of HFC134a to levels of 350,000 t yr^{-1} in 2025.

2.1.8
Pentafluoroethane (CHF_2CF_3/HFC125)

Pentafluoroethane (HFC125) is under way to reach considerable commercial relevance as a major constituent of new HFC blends (see Table 2) for stationary cooling applications. There are so far no atmospheric observations of HFC125.

2.1.9
Hexafluoropropane ($CF_3CH_2CF_3$/HFC236fa)

Hexafluoropropane may become one of the main compounds for the use as blowing agent for the production of closed cell foams [14] until halocarbon-free technologies become more widely accepted. Atmospheric observations of HFC236fa are currently not available.

2.1.10
Heptafluoropropane (CF_3CHFCF_3/HFC227ea)

This substance is proposed and is beginning to be used as a physiologically safe as well environmentally more acceptable aerosol propellant for pharmaceutical sprays [26]. Whether it will be widely used as a propellant for aerosols in other applications remains uncertain, as hydrocarbons and to some degree HFC134a and HFC152a have already reached a very competitive position in this field. HFC227ea also appears to be a likely candidate to substitute the brominated halons (CF_2ClBr, CF_3Br and $C_2F_4Br_2$) in fire extinguishing [14]. There are so far no atmospheric observations of HFC227ea.

2.2 Perfluorocarbons

This class of fluorocarbons is characterized by its unique stability. The substances are inflammable and in general non-toxic. Perfluoroisobutene ($(CF_3)_2CCF_2$) is an important exception with very high toxicity. The stability of perfluorocarbons lends them extremely long atmospheric lifetimes of several thousand years (Table 1). Moreover perfluorocarbon molecules strongly absorb infrared radiation. They are potent greenhouse gases due to these radiative properties and their stability. Owing their relatively low atmospheric abundance perfluorocarbons currently contribute less than 0.4% to the total anthropogenic radiative climate forcing.

2.2.1 *Tetrafluoromethane* (CF_4/FC14)

Atmospheric tetrafluoromethane was first discovered by Gassmann [27] who detected traces of CF_4 in highly pure krypton from air-liquefaction. Unaware of significant anthropogenic sources he speculated on natural sources [28]. Sampling the exhaust plume of an aluminum smelter onboard of an airplane Penkett et al. [29] could show that the electrolytic production of primary aluminum is a major source of atmospheric CF_4 and C_2F_6. This confirmed older emission measurements by Henry and Holiday [30] who had found CF_4 and C_2F_6 in the off-gas from an electrolysis cell during the so called anode-effect. The anode-effect is an intermittent phase of alumina-depletion of the electrolyte. Searching for possible natural sources of CF_4 Rasmussen et al. [31] analyzed air from volcanic fumeroles on Hawaii. No significant volcanic contribution to the atmospheric reservoir of CF_4 was detected. It was therefore assumed that CF_4 lacks significant natural sources. Interpreting their stratospheric measurements of CF_4 and C_2F_6 with a 1-dimensional model of global vertical atmospheric transport Fabian et al. [32] suggested the existence of a pre-industrial atmospheric background of about 65% of the 1985 atmospheric burden of CF_4. A first assessment of the atmospheric fate of CF_4 was carried out by Cicerone [33]. For CF_4 he proposed the extraordinary long atmospheric lifetime of greater 50,000 yr, which more than a decade later was confirmed by Ravishankara et al. [34]. Wang and co-workers [35] were the first to quantify the contribution of atmospheric CF_4 to the greenhouse-effect. In the 1990s it was this radiative property of CF_4 that led to an increase of public concern about the effects of the accumulation of CF_4 in the earth's atmosphere [36, 37]. A recent estimate of the atmospheric relevance of current emissions of CF_4 in comparison to carbon dioxide (CO_2) was given by Weston [38]. Harnisch et al. [39] lately showed that

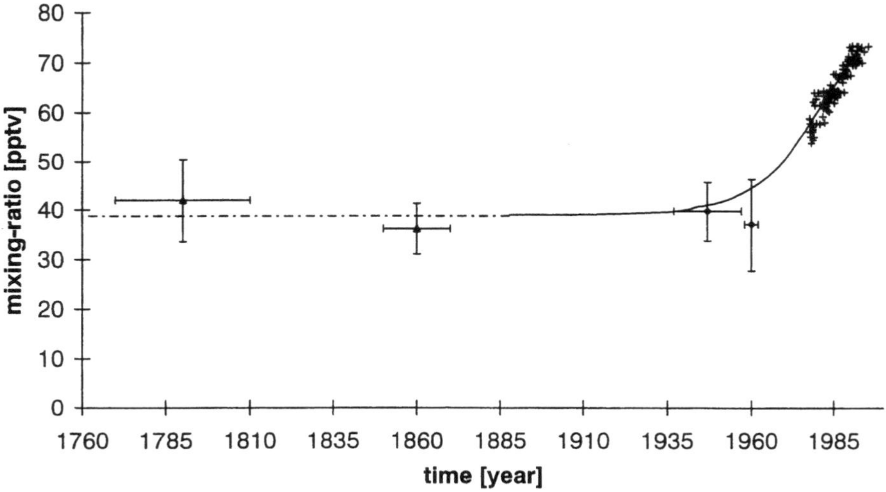

Fig. 4. The atmospheric chronology of tetrafluoromethane. After [39]

natural sources contributed about half of the current atmospheric reservoir of CF_4 (Fig. 4). They presented evidence for the relevance of degassing-processes from the Earth's crust, where CF_4 appears to have its radiogenic origin in fluoride-minerals [40]. CF_4 is also used in plasma etching processes of semiconductor production. This application probably contributes less than 300 t yr^{-1} [41] to the total increase of 15,000 t yr^{-1} of the atmospheric reservoir of CF_4 [89]. The lion's share of these emissions must be attributed to aluminum production [42]. Manufacturers of aluminium have recently carried out a worldwide survey to estimate emissions of CF_4 and C_2F_6 from smelters [43] showing some progress in the reduction of specific emissions between 1990–1993. Eventually emissions of CF_4 and C_2F_6 and many other pollutants from aluminum smelters could be eliminated through the use of inert anode-materials which are currently in an early experimental stage of development [44].

2.2.2
Hexafluoroethane (C_2F_6/FC116)

While the presence of CF_4 in the Earth atmosphere was discovered by chance in the early 1970s [27], it was part of a systematic search that the tropospheric abundance of the homologous C_2F_6 was first determined [45] in 1981. Since then its average mixing-ratio has linearly increased from 1.2 to 2.6 pptv in 1995 [89]. This increase corresponds to an annual injection of about 2,000 t yr^{-1}. This injection can be almost completely attributed to the production of primary aluminum. Its use as a source of fluorine in plasma-etching in the production of semiconductors is limited to about 200 t yr^{-1} [46]. Natural sources have not been identified so far [40].

2.2.3
Octofluoropentane (C_3F_8/FC218)

Octofluoropentane currently has very limited economic relevance. There are some applications for plasma etching in semiconductor production [47]. In the order of 100 t yr^{-1} FC218 is emitted during the production of primary aluminum [40]. There are currently no atmospheric measurements of C_3F_8 available. Unless further relevant sources exist, its atmospheric abundance should remain below 0.2 pptv for a long time.

2.2.4
Octocyclobutane (c-C_4F8/RC218)

Octocyclobutane currently has very limited economic relevance. There are some applications for plasma etching in semiconductor production [48]. For the year 1994 Sturges et al. [49] report a tropospheric mixing-ratio of 0.4 pptv decreasing to a value of 0.1 pptv at altitudes of 25 km. Thus in 1994 the atmospheric reservoir of c-C_4F_8 had a mass of about 10,000 t. The vertical gradient indicates a magnitude of global emissions of 1,000–2,000 t yr^{-1} if an infinite atmospheric lifetime is assumed as a first approximation. The nature of the source of these emissions is unclear. c-C_4F_8 is formed through the dimerization of tetrafluoroethene [3]. Possibly it is the combustion or thermal decomposition of polytetrafluoroethylene (PTFE) and other fluoropolymers [50] that is responsible for the accumulation of c-C_4F_8 in the atmosphere. The annual consumption of fluoropolymers in Western-Europe in 1994 amounted to 16,000 t yr^{-1} [51]. Worldwide production, use and disposal of fluoropolymers could therefore well suffice for the unexplained emissions.

2.2.5
Perfluoroalkylcycloalkanes

This group of substances has very little commercial relevance. Because of their low atmospheric abundance ($0.1-30 \times 10^{-15}$ mol mol^{-1}) and their excellent detectability they can be very effectively used as tracers of atmospheric motion if purposely injected into the atmosphere [52–54]. For single experiments amounts of 0.1–10 kg of chemicals like $C_6F_{11}CF_3$-perfluoromethylcyclohexane (PMCH), $C_5F_9CF_3$-perfluoromethylcyclopentane (PMCP) or $C_6F_{10}(CF_3)_2$-perfluorodimethylhexane (PDCH) are applied. Their respective calculated atmospheric reservoirs in the mid-1980s were in the order of 100–2,000 t with unclear origins.

2.3
Further Organic Fluorine Compounds

This paragraph covers a group of substances that are only loosely related to each other. The particular classes were selected somewhat arbitrarily. The reader may refer to [3] for information on further anthropogenic fluorine compounds and to [55] for information on further natural fluorine compounds.

2.3.1
Fluoroacetic Acids

Three different substances belong to this group: monofluoroacetic acid (MFA) (CH_2FCOOH), difluoroacetic acid (DFA) (CHF_2COOH) and trifluoroacetic acid (TFA) (CF_3COOH). All three of them are more or less toxic. MFA can be found in many poisonous tropical plants [56]. Mimicking acetate in the metabolism it can be extremely harmful to mammals. TFA on the other hand is believed to be a purely anthropogenic compound, which is formed as a degradation product of fluoroethanes containing the CF_3-group [18, 19]. Recent measurements of TFA in ambient air (50 pg m^{-3}/0.01 pptv), rain (100 ng l^{-1}) and surface waters (200 ng l^{-1}) showed that TFA is an ubiquitous atmospheric compound [57, 58]. A first global extrapolation from these data yields an annual deposition of some 10,000 t yr^{-1} of TFA to the surface. The only known atmospheric precursors of TFA are halothane ($CHClBrCF_3$), tetrafluoroethane (HFC134a), trifluorochloroethane (HFC123), and tetrafluorochloroethane (HCFC124). The annual release of these substances into the atmosphere does not seem to fully explain the origin of the TFA concentrations measured. There is some evidence that TFA and similar compounds may form through the thermal combustion of fluoropolymers [59]. The potential future accumulation of TFA from the degradation of hydrofluoroethanes has been discussed by Tromp et al. [60]. These calculations assumed a complete lack of environmental sinks for TFA, which may not be true as Visscher et al. claim to have observed microbial degradation of TFA in anoxic and oxic sediments [61]. This finding has yet not been confirmed [62]. The environmental fate of TFA remains subject to intensive research [63].

2.3.2
Carbonyl Fluorides ($COClF$, COF_2, $COHF$, CF_3COF, CF_3COCl)

The carbonyl fluorides $COClF$, COF_2, $COHF$, CF_3COF, and CF_3COCl are degradation products of common halocarbons [64]. $COClF$ has been detected above the tropopause with about 4 pptv in 1986/87 using a matrix isolation Fourier transform infrared spectrometer technique [65]. Model calculations for 1989 predict mixing-ratios of more than 100 pptv for 30 km altitude sharply decreasing above due to photolysis [104]. In 1992 Zander et al. spectroscopically detected COF_2 in the stratosphere [66]. They found a maximum mixing-ratio of about 200 pptv in the altitude range of 30–40 km in good agreement with model results [104]. The other carbonyl fluorides listed above have as yet not been observed in the atmosphere. For the year 2020 Kanakidou et al. [64] predict as maximum mixing-ratios: $COClF < 1$ pptv, $COF_2 < 5$ pptv, $COHF < 120$ pptv, $CF_3COF < 1$ pptv, $CF_3COCl < 1$ pptv. COF_2 and $COHF$ would thus be the only substances to build up a relevant tropospheric abundance. Nevertheless, it should be kept in mind, that future source gas abundance as well as kinetic data on which these calculations are made are quite uncertain and may need future verification through atmospheric measurements.

2.3.3
Trifluoromethanol (CF_3OH)

CF_3OH is formed in the atmospheric degradation of compounds containing the CF_3-group. Its likely atmospheric fate was recently investigated [67, 68]. Its lifetime in the troposphere as well as in the stratosphere is likely to be controlled by heterogeneous processes on the surfaces of aerosol particles. For the troposphere the atmospheric lifetime is determined to be about 2 days [68] in comparison to a value of about 5 years in the aerosol-poor stratosphere [67]. Atmospheric observations of CF_3OH have not been reported so far.

2.3.4
Tetrafluoroethene (C_2F_4)

Tetrafluoroethene as the monomer for the production of PTFE is of great commercial importance. There are no data available on losses into the atmosphere occurring during its production and handling. Because of the double bond, C_2F_4 is expected to undergo rapid photochemical degradation. Atmospheric measurements of C_2F_4 have not been reported with the exception of Stoiber et al. who claim to have detected C_2F_4 in volcanic gases at Santiaguito in Guatemala [69].

2.3.5
Fluorinated Pesticides

There is a group of commonly used pesticides into which a CF_3-group is introduced in order to increase the stability and hydrophobic character of these compounds [3]. Examples of commercial relevance are the herbicides trifluralin (CAS-Reg. No.:1582-09-8), profluralin (CAS-Reg. No.: 26399-36-0), benfluralin (CAS-Reg. No.: 1861-40-1), fluodifen (CAS-Reg. No.: 15457-05-3), acifluorfen (CAS-Reg. No.: 50594-66-6), fluometuron (CAS-Reg. No.: 2164-17-2), or the insecticide fluvalinate (CAS-Reg. No.: 69409-94-5). It is difficult to estimate the commercial relevance of this group of substances. Trifluralin in 1995 was produced at about 15,000 t yr^{-1} [70]. A number of 35,000–40,000 t yr^{-1} for the total worldwide production of fluorobenzenes in 1985 [3] is available. Field volatilization losses into the atmosphere of up to 90% for trifluralin may occur depending on the soil type and the application methods [71, 72]. Atmospheric trifluralin then undergoes a rapid photo-transformation within few hours [73, 74]. The first steps of the environmental degradation of trifluralin are quite well understood whereas final degradation products are unknown [75–77]. The environmental fate of the pesticides listed above has remained quite unclear despite their large scale use.

Table 4. Fluorinated anesthetics and their tropospheric lifetimes [80]

Substance	Formula	Troposph. Lifetime
halothane	$CHClBrCF_3$	2 years
enflurane	$CHClFCF_2OCHF_2$	6 years
isoflurane	$CF_3CHClOCHF_2$	5 years
desflurane	$CF_2HOCFHCF_3$	unknown
sevoflurane	$CFH_2OCH(CF_3)_2$	1.4 years

2.3.6
Anesthetics

A group of fluorinated substances is used as volatile anesthetics with a world consumption in the order of 100–1,000 t yr^{-1} each [80]. Their tropospheric lifetimes are in the order of years (Table 4) [80]. For none of the gases listed in Table 4 a tropospheric mixing-ratio higher than 0.3 pptv can be expected for a worldwide release of 1,000 t yr^{-1}. Their stratospheric impact is thus very limited. Halothane is the only fluorinated anesthetic for which stratospheric profiles have been reported [49]. Though lacking absolute calibrations these measurements show that in the stratosphere 90% of the tropospheric abundance of halothane is lost in photochemical processes at altitudes below 20 km releasing active bromine that effectively participates in the catalytic destruction of ozone. The future market of fluorinated volatile anesthetics is unlikely to grow rapidly despite some potential in the developing world. Intravenous narcosis today appears to be favored in many applications. With growing awareness about the protection of the ozone-layer the commercial relevance of halothane should gradually decrease despite its attractive price.

2.3.7
Fluorinated Dibenzodioxines and Dibenzofurans

Very little is known about the occurrence, sources and toxicity of fluorinated dibenzodioxines and -furans which are homologous to their highly toxic chlorinated correspondents which aroused widespread public concern. Recently, a first assessment has become available [78] showing that fluorinated dibenzodioxines and -furanes can be formed in considerable amounts as thermal degradation products of CFCs.

2.4
Long-lived Inorganic Fluorine Compounds

2.4.1
Sulfur Hexafluoride (SF_6)

Sulfur hexafluoride is a very long-lived constituent of the atmosphere. The SF_6 molecule is the most potent greenhouse gas of commercial relevance [79].

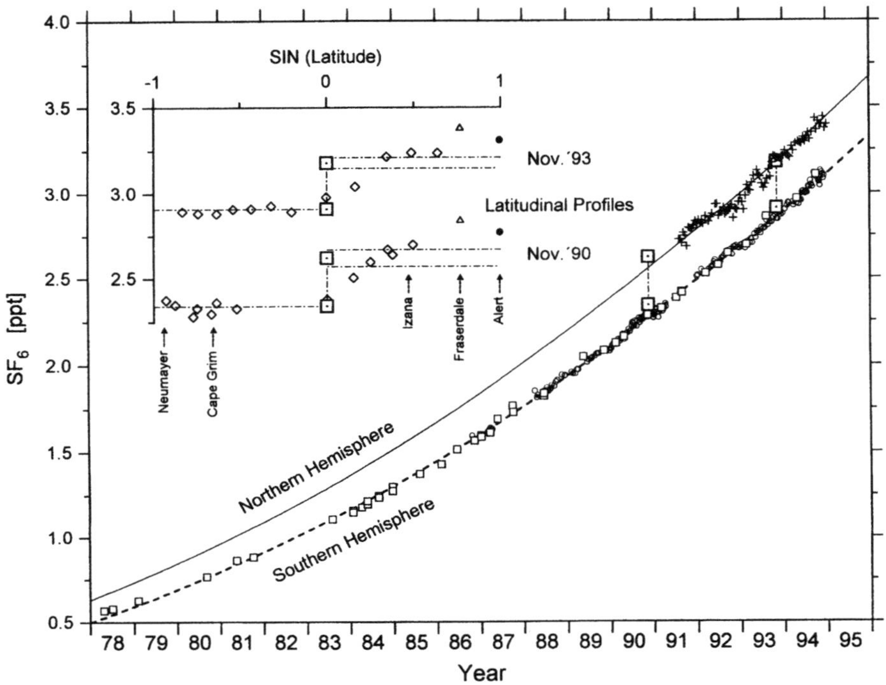

Fig. 5. The atmospheric chronology of sulfur hexafluoride. After [83]

Atmospheric SF_6 is believed to be mainly anthropogenic even though SF_6 has been extracted from fluorspar minerals [39]. The presence of SF_6 in the atmosphere was first discovered by Lovelock who reported a tropospheric background abundance of 0.03 pptv for 1970 [80]. Stratospheric measurements carried out by Krey et al. [81] show that already in 1970 a tropospheric mixing-ratio of 0.15 pptv was present. Regardless of this discrepancy it is evident that its tropospheric abundance has rapidly increased since then. The atmospheric chronology of SF_6 from the late 1970s to the present and its meridional distribution have been established through the work of Maiss, Levin and co-workers [82, 83] (Fig. 5). Estimates of the world production of SF_6 are in a range of 5,000–8,000 t yr^{-1} [83]. SF_6 is mainly used in electrical high voltage switching equipment and as a protection gas for magnesium casting [37]. There is currently a large discrepancy of several thousand tons between reported emissions and the observed increase of the atmospheric abundance of SF_6 [83, 84] which currently amounts to 6,000 t yr^{-1}. The well known atmospheric history of SF_6 proves to be extremely valuable for studies of the exchange of water masses inside lakes [85, 86], for studies of the exchange between the atmosphere and the oceans [87], between air and the firn-ice-system of glaciers [88], or between the troposphere and the stratosphere [89].

2.4.2
Nitrogen Trifluoride (NF_3)

No measurements of the concentrations of atmospheric nitrogen trifluoride have yet been reported. It is used in plasma etching processes of semiconductor production [90]. If NF_3 accumulated in the atmosphere, the amounts produced worldwide would lead to global mixing-ratios of below 0.05 pptv [91].

2.5
Short-lived Inorganic Fluorine Compounds

The research interest in these substances mainly originates from their acute and chronic toxic effects on plants, animals and humans [92]. In the troposphere these fluorine compounds can be easily removed from the atmosphere through wet or dry deposition to surfaces. For analytical reasons in most cases a discrimination between any of the compounds HF, SiF_4, H_2SiF_6, F_2, Na_3AlF_6, AlF_3, CaF_2, NaF or $Ca_5(PO_4)_3F$ was neither carried out for emission measurements nor for the determination of ambient concentrations. It is therefore currently impossible to individually quantify the tropospheric abundance of these compounds. Frequently the term "fluoride" is used to refer to inorganic fluorine compounds that dissociate into ions in water. In the text below this terminology will also be used.

During the last decades it was attempted to determine a global background concentration of fluoride in precipitation of unpolluted regions, aiming to derive information on the relative global significance of the major large natural sources of fluoride: volcanoes, terrigenous dust, sea-spray and biomass-burning (Fig. 6). One important limit to this approach stems from the short atmospheric

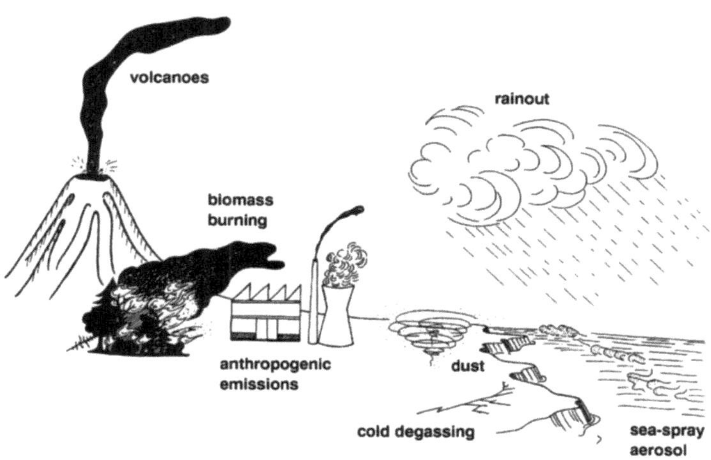

Fig. 6. Processes that determine the global budget of atmospheric fluorine. With courtesy C. Mascher, Göttingen

residence times of fluoride in the atmosphere. Wet deposition is a fast removal process of fluoride from the atmosphere. Even though the global distribution of inorganic fluoride has not been modeled yet one may expect it to show similar patterns as that of HCl [93], SO_2 and SO_4^{2-} [94]. Three-dimensional (10° × 10°) models predict distributions of SO_2 and SO_4^{2-} concentrations two to three orders of magnitude higher for industrialized countries of the north than for the remote areas of the south [94]. A similar situation can also be expected for the inorganic fluorine compounds. While explicit average turnover-times for atmospheric hydrogen flouride (HF), SiF_4, H_2SiF_6 and dissolved fluorides have not been reported, the values of 1–2 days of SO_2 [94] for the former three and 3–6 days of SO_4^{2-} [94] for the later appear to be reasonable estimates. A recent estimate for the average residence time of HCl is 36 hours in the boundary layer of the atmosphere and about two month for the free troposphere [93]. Recent measurements of fluoride in precipitation in southern Norway [95] and glacial firn and ice from Greenland [119] clearly confirm the strong influence of regional sources of fluoride.

As emissions of fluorides are deposited regionally it is of very limited use to calculate a global source strength for the substances listed above. Such a global analysis of fluorine flows into and out of the atmosphere will nevertheless be carried out below in order to allow a comparison with organic fluorine compounds. It is rather difficult to obtain a reasonable estimate of annual natural and anthropogenic emissions for the fluorides listed above. Due to the short atmospheric residence-time of fluorides, one has to rely exclusively on emission measurements as, unfortunately, globally representative measurements of fluoride in precipitation are not available.

2.5.1 *Hydrogen Fluoride* (HF)

Hydrogen fluoride has been known as an atmospheric pollutant for more than 100 years, when injury of plants was reported from the vicinity of super-phosphate fertilizer industries in Germany [96]. Since then many incidents involving fluoride have occurred close to industrial sources like aluminum smelters, brick factories, and enterprises involving the mining and processing of phosphates. The combustion of coal [97, 98] and the incineration of wastes [97] are two further important anthropogenic sources of HF. In the western world stringent emission controls have greatly reduced the toxic effects on plants and animals in the direct vicinity of such enterprises. The most important natural source of HF is volcanism [98, 99]. Biomass burning [100] as well as the release of HF from sea-salt aerosol [100, 101] appear to contribute insignificantly (see Sect. 2.6.1).

For the stratosphere where HF forms as a degradation product of the large range of different fluorohalocarbons optical [102] as well as mass-spectrometric [103] measurements are available. These data show that HF mixing-ratios increase with altitude from below 10 pptv near the tropopause to about 1 ppbv above 30 km altitude. This is consistent with modeled stratospheric distributions [104] and reflects the high efficiency of photochemical destruction of fluorohalocarbons and carbonyl fluorides. The abundance of HF in the undisturbed stratosphere was probably orders of magnitude lower as relevant organic source compounds except for the inert CF_4 were virtually absent. The exchange of air masses between troposphere and stratosphere [105] and the direct injection of hot volcanic gases from large eruptions [99] must have determined the stratospheric HF abundance. The free tropospheric mixing-ratio of HF of few part-per-trillion or less thus places an upper limit on the natural abundance of stratospheric HF.

2.5.2
Silicon Tetrafluoride (SiF_4)

No measurements of ambient concentrations of atmospheric SiF_4 have been reported so far. Using an optical remote sensing technique Francis and coworkers [106] could recently show that, depending on the temperature of gases, SiF_4 can be a major fluorine compound in volcanic gas vents. They found mass-ratios of SiF_4 to HF of 0.1 for fumeroles of 200–450°C temperature at Mt. Etna and Volcano (Italy). This is in reasonable agreement with model-calculations by Symonds et al. [107]. Graedel et al. [108] list various industrial sources which emit SiF_4: e.g. the production of aluminum, fertilizer or phosphoric acid. Currently there is no quantitative information available on the fate of atmospheric SiF_4. Like HF it will probably be removed quickly from the atmosphere through wet deposition and occurring as fluoride in precipitation. Its concentration in the remote atmosphere is probably much lower than that of HF.

2.5.3
Fluorine (F_2)

Fluorine (F_2) is produced electrolytically with about 17,000 t yr^{-1} [1]. It is mainly consumed for the production of uranium hexafluoride (UF_6) and sulfur hexafluoride (SF_6). Losses to the environment are strongly restricted because of its high toxicity. Measurements of atmospheric concentrations of fluorine are unavailable. Its mixing-ratios can be expected to be extremely low due to its high reactivity. For the same reason natural emissions of gaseous F_2 must be considered extremely unlikely even though its existence inside the crystal matrix of some types of fluorspar is established [109].

2.5.4
Hydrofluorosilic Acid (H_2SiF_6)

In the USA hydrofluorosilic acid is produced commercially at a scale of 100,000 t yr^{-1} [2]. In many cases it is a by-product from phosphate and superphosphate plants and the production of HF and aluminum fluoride. Despite stringent pollution control H_2SiF_6 is released into the environment. Quantitative information on emissions and concentrations in ambient air are lacking. Equilibrium calculations indicate that H_2SiF_6 is released naturally from volcanoes [107].

2.5.5
Particulate Fluoride

For a complete view on atmospheric fluorine one also has to consider particulate fluoride, which makes up a considerable fraction of tropospheric fluorine in natural as well as in industrial sources. Particulate fluorides are emitted from industrial sources mainly as cryolite (Na_3AlF_6), aluminum fluoride (AlF_3), fluorspar (CaF_2), sodium fluoride (NaF) or apatite ($Ca_5(PO_4)_3F$). Grjotheim et al. [110, 111] have presented a very thorough analysis of the distribution of fluoride-compounds for the emissions of primary aluminum production. Model calculations for Mount St. Helens by Symonds and Reed [107] indicate that volcanoes also emit these compounds as particles. A large natural fluorine input into the troposphere through terrigenous dust was proposed by Bewers and Hayson [112]. Recently the global source strength for natural mineral dust was estimated to be in the range of $200-5000 \times 10^{12}$ g yr^{-1} [113, 114]. The spatial and temporal distribution of sources and sinks of atmospheric dust is highly variable [114]. The actual strength of the global source of mineral dust and the average fluorine content of soils and surface minerals (approx. 200–500 ppm) together determine the natural dust contribution to the deposition of fluoride. The global natural deposition of particulate fluoride is therefore uncertain within two orders of magnitude ($4 \times 10^{10} - 3 \times 10^{12}$ g yr^{-1}), which makes a comparison with anthropogenic emissions difficult. In arid areas natural dust will generally be the dominant source of particulate fluoride albeit mainly being deposited to similar soils it originated from and thus leaving only a small share to long range transport. The relevance of an oceanic source of fluorine from sea-salt aerosol has been extensively discussed [115–118] and must today be considered to be negative as dust particles are deposited over the oceans [100, 119, 120]. Recently it was estimated that about 1.7×10^{14} g yr^{-1} of sea-salt aerosol are exported from the oceans to the continents [120]. This value corresponds to a fluorine deposition of only 2×10^8 g yr^{-1} for an average fluorine content of 1.3 ppm in ocean water [122]. If also sea-salt aerosol is considered that is deposited back to the oceans (1.1×10^{16} g yr^{-1} [122]) an emission of 1.4×10^{10} g yr^{-1} of fluorine bound to sea-salt aerosol is calculated which is at the lower end of the likely estimates of contributions from volcanoes and terrigenous dust.

2.6
Natural and Anthropogenic Emissions of Fluorides

The relative relevance of anthropogenic emissions of short-lived gaseous and particulate atmospheric fluorides in comparison to natural sources is still rather unclear. Here it will be attempted to critically review the sparse available information.

2.6.1
Natural Emissions of Fluorides

A positive contribution of fluoride from the oceans can today be ruled out [117, 119]. Terrigenous dust was proposed as a large natural source of atmospheric fluoride [112]. In the paragraph on particulate fluoride (Sect. 2.5.5) the contributions of the oceans as well as of terrigenous dust to the global cycle of fluorine were discussed more extensively. The best understood natural source of atmospheric fluoride is volcanism. Even in this case estimates of emissions of 0.4×10^{12} g yr^{-1} [98] or $0.06-6 \times 10^{12}$ g yr^{-1} [99] are uncertain within two orders of magnitude. The upper estimate must be considered somewhat unrealistic as it assumes equal global volcanic emissions of SO_2, HCl and HF on a molar basis. Usually molar ratios HCl/HF of 3–200 are found in volcanic gases [98, 121]. A likely range of $0.01-1 \times 10^{12}$ g yr^{-1} of fluoride appears to be a more reasonable estimate for a year of moderate volcanic activity. Biomass burning was also proposed as a possible major source of atmospheric fluoride [100]. As measurements remain unavailable, an upper limit of 4×10^9 g yr^{-1} can be calculated if the total fluorine content of typically about 0.5 ppm for land vegetation [122] is released from the 7×10^{15} g yr^{-1} of dry biomass estimated to be currently burned annually [123]. This quantity is small compared to the volcanic contribution and must be corrected to natural levels of biomass burning which are believed to be only 5% of the present extent [124]. Biomass burning is thus not responsible for the variability of the fluoride-content observed in Greenland ice [119]. It is more likely that changes in the precipitation and atmospheric circulation pattern in the northern hemisphere have episodically enhanced the deposition of volcanic fluoride from Iceland and mineral dust from North America.

2.6.2
Anthropogenic Emissions of Gaseous Fluorides

Individual industrial sources of fluoride were discussed above. As reliable emission factors for relevant processes are lacking, a different approach is chosen here to obtain a reasonable estimate for the order of magnitude of anthropogenic fluoride emissions. First, the total amount of inorganic fluorine industrially processed is calculated. Then an appropriate percentage is assumed that is released into the environment via the gas-phase.

Table 5 shows the results of these rough estimates. For the present, a fluoride turnover through thermal and chemical processes, of 1×10^{13} g yr^{-1}, is estimated. This is in reasonable agreement with older figures [115] but somewhat higher due to the larger number of processes included into the evaluation. Emission-factors were retrieved from the data-bank of EPA [97]. These data indicate that for most processes in the USA today the ratio between fluorine emitted into the atmosphere and fluorine processed is well below 1%. This value provides a reasonable lower limit for worldwide inorganic fluorine emissions yielding annual emissions of at least 0.1×10^{12} g yr^{-1}. An upper limit for the worldwide release of fluorine from the industry can be obtained from the review by Smith and Hodge [92] who report a release factor of about 30% for the USA in the late 1960s. From Table 5 the author of this article obtains a likely range of $0.3-1 \times 10^{12}$ g yr^{-1} for the global annual anthropogenic emissions of fluoride.

Table 5. Inorganic fluorine handled and processed industrially and estimated emissions into the atmosphere. The numbers of this Table are rough estimates only as more reliable data are unavailable

Process	Annual activity [g yr^{-1}]	Spec. F turnover [g g^{-1}]	F turnover [g yr^{-1}]	Estimate of relative and absolute [g] gaseous emissions of F
aluminium	20×10^{12} [111]	0.01–0.02 [111]	$0.2–0.4 \times 10^{12}$	$\leq 20\%/0.06 \times 10^{12}$
phosphate	40×10^{12} [133]	0.08	3×10^{12}	$\leq 5\%/0.15 \times 10^{12}$
cement	1300×10^{12} [134]	5×10^{-4}	0.7×10^{12}	$\leq 30\%/0.2 \times 10^{12}$
combustion of coal	3000×10^{12} [98]	1×10^{-4} [98]	0.3×10^{12}	$\leq 30\%/0.1 \times 10^{12}$
hydrofluoric acid	1.2×10^{12} [2]	0.95	1.1×10^{12}	$\leq 5\%/0.05 \times 10^{12}$
steelmaking	700×10^{12} [134]	0.002–0.01 [2, 135]	3×10^{12}	$\leq 20\%/0.6 \times 10^{12}$
TOTAL			10×10^{12}	$\leq 1 \times 10^{12}$

Table 6. Estimated emissions, residence and removal of atmospheric fluorine (F) in 1992

	Annual emissions [10^9 g(F) yr^{-1}]	Residence time [years]	Atmospheric reservoir [10^9 g(F)]	Annual removal [10^9 g(F) yr^{-1}]
Fluoride from natural mineral dust	40–3000	0.01	0.4–30	40–3000
Inorganic fluorides from volcanoes	10–1000	0.01	0.1–10	10–1000
Sea-spray aerosol & biomass burning	< 20	0.01	< 0.2	< 20
Anthropogenic short-lived fluorides	300–1000	0.01	3–10	300–1000
Hydro(chloro) fluorocarbons	120	15	800	50
Chlorofluorocarbons	250	80	5000	65
Perfluorinated compounds	20	> 5000	1000	< 0.2

3 Aspects of Fluorine Geochemistry

A rather simple approach to characterize the atmospheric geochemistry of fluorine is chosen here. It will take into account the annual global emissions, the residence times and the sizes of the atmospheric reservoirs and the annual removal rate for classes of species. Rough estimates for seven major groups of fluorine compounds are given in Table 6.

The group of short-lived inorganic fluorine compounds has a high annual turnover in terms of emissions and removal, while exhibiting only a rather small atmospheric reservoir at any given moment. The other entries of Table 6 show a different behavior. Because of their long atmospheric lifetimes the other three groups have not reached atmospheric equilibrium. Their emissions exceed losses leading to a continuous increase of the size of the atmospheric reservoir. The total atmospheric reservoir of fluorine can be estimated to about 7×10^{12} g for the year 1995. This is more than a factor of ten higher in comparison to the natural situation, when 0.5×10^{12} g were contributed through CF_4 and less than 0.05×10^{12} g through gaseous and particulate inorganic fluorides.

The size of this atmospheric reservoir (natural, 1995 and 2050) of fluorine can be compared with that of its other geochemical reservoirs (Fig. 7) (see also [125]). The total masses of fluorine contained in the atmosphere, in the biosphere and the cryosphere and the freshwaters agree within one order of magnitude. From groundwater, to the soils, to the oceans and finally to the crust the mass of the respective reservoirs of fluorine increases dramatically. This is partly due to the higher total masses of these spheres. The large variations of the mass contents of fluorine between these spheres are the important second factor (Fig. 8). The concentration of an element in a reservoir is driving the chemical processes. Fluxes from one reservoir to another in comparison to the reservoir

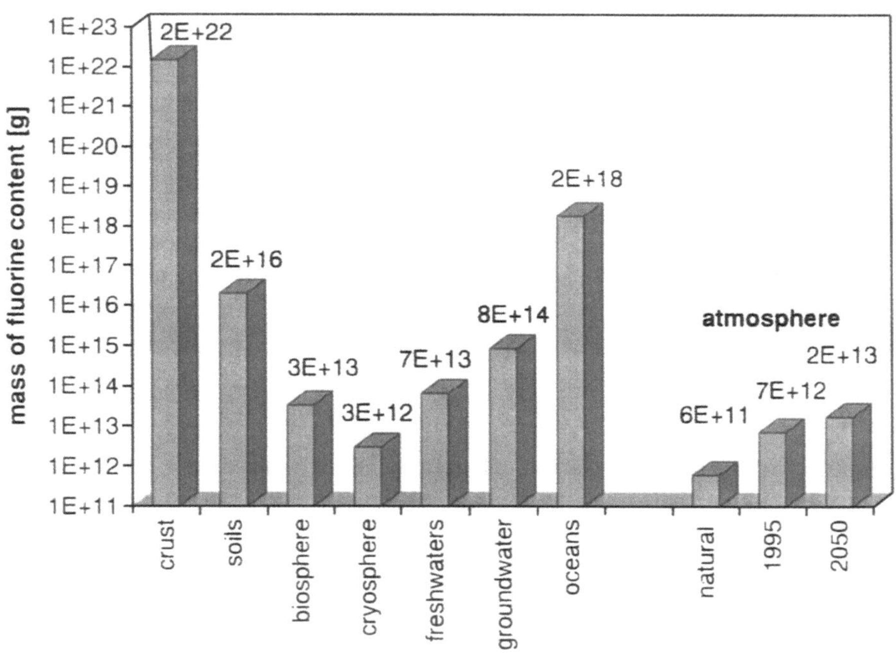

Fig. 7. Masses of geochemical reservoirs of fluorine using data from [15, 119, 122, 138]. For the calculation of the atmospheric reservoir of 2050 the IPCC scenario IS92de [126] was used and constant emissions of CF_4 of 15,000 t yr^{-1} [89] were assumed

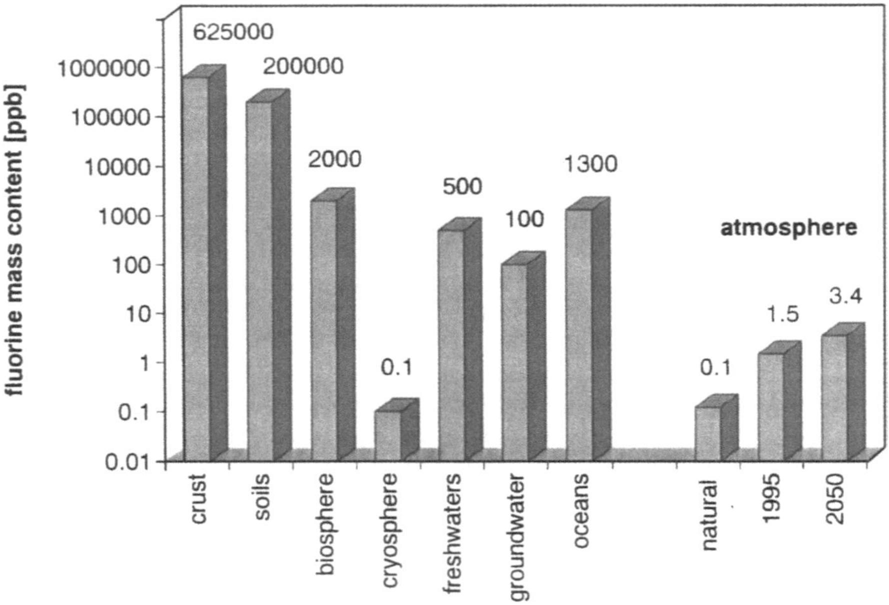

Fig. 8. Mass contents of fluorine for different spheres using data from [115, 119, 122, 138]. For the calculation of the atmospheric mass content of fluorine in 2050 the IPCC scenario IS92de [126] was used and constant emissions of CF_4 of 15,000 t yr^{-1} [89] were assumed

size can be used to obtain a rough measure of the probability of a relevant chemical disturbance of the reservoir-composition. A simple analysis as presented in this chapter of course should not be over stretched. Nevertheless, it is evident that safety margins are small in the case of fluorine if anthropogenic fluxes, their natural equivalents and the sizes of relevant element reservoirs are taken into account.

4 Concluding Remarks

Annual anthropogenic emissions of fluorine compounds today have reached and possibly exceeded the magnitude of natural sources. The atmospheric reservoir of fluorine has increased tenfold through industrial emissions of long-lived fluoro-organic compounds. The amount of HF in the stratosphere has grown by orders of magnitude as a consequence of the use of fluorohalocarbons. These changes may appear dramatic in the light of a geochemical substance-flow analysis.

The actual and potential environmental effects of these changes of the atmospheric composition are still partly uncertain. According to current knowledge fluorine compounds do not play a significant role in any major air-chemical process. The contribution of HFCs and perfluorinated compounds to the increase of the greenhouse effect could in the future reach and exceed the present level of the CFCs. So far there is only limited evidence that degradation products of HFCs will have adverse short-term effects on the environment. The local and regional toxic effects of inorganic fluoride emissions are sufficiently well understood. While fluoride emissions are under stringent control in most industrialized countries, there are still large industrial emissions with devastating regional effects in other parts of the world.

Three problematic aspects of the environmental behavior of organic fluorine compounds should be emphasized. First, their long atmospheric lifetimes permit a buildup of high atmospheric concentrations which would significantly contribute to global warming and could potentially lead to chemical perturbations of the stratosphere. Secondly, some atmospheric degradation products of organic fluorine compounds do not seem to be metabolized easily and innocuously in organisms. Thirdly, the natural fluorine content of the atmosphere has already multiplied through anthropogenic activities. For these three reasons organic fluorine compounds should remain interim-solutions in their large scale applications until technical alternatives to the use of organic fluorine compounds become readily available.

The technical processes that lead to a release of inorganic and organic fluorine compounds into the atmosphere are threefold significant: in terms of their benefits for everyday-life, in terms of their economic relevance as well as in respect to their known and potential environmental effects. The scientific understanding of the origin and fate of atmospheric fluorine compounds is still quite fragmentary. In order to obtain a sound foundation for political decisions on future environmental regulation a more systematic research on environmental substance flows is overdue.

Acknowledgements. The author would like to thank A. Klein from the University of Bayreuth for the latest information on environmental TFA and its precursors. Many thanks also to M. Buchert from the Öko-Institut in Darmstadt and H. Dölling from VAW Aluminium AG in Bonn for many fruitful discussions on fluorinated compounds in the environment. A. Leisewitz and W. Schwarz from Öko-Recherche in Frankfurt provided an early version of their manuscript on emissions of fluoro-carbons in Germany. M. Maiss from the Max-Planck-Institut für Chemie kindly contributed a diagram of his atmospheric chronology for sulfur hexafluoride. C. Mascher from Göttingen prepared the illustration of the global atmospheric cycle of fluorine. Many thanks also to G. Wieland from Göttingen for her steady encouragement. The author also gives credit to the Climate Monitoring and Diagnostics Laboratory (CMDL) of the US-National Oceanic and Atmospheric Administration (NOAA) in Boulder for providing the data of their global measurements of HFC 134a via the World Wide Web.

5
References

1. Ullmann's Encyclopedia of Industrial Chemistry; Fluorine, 5th edition, Vol **A11**, pp 293 - 305, VCH Verlagsgesellschaft, Weinheim, 1988
2. Ullmann's Encyclopedia of Industrial Chemistry; Fluorine Compounds, Inorganic, 5th edition, Vol **A11**, pp 307 - 348, VCH Verlagsgesellschaft, Weinheim, 1988
3. Ullmann's Encyclopedia of Industrial Chemistry; Fluorine Compounds, Organic, 5th edition, Vol **A11**, pp 349 - 392, VCH Verlagsgesellschaft, Weinheim, 1988

4. Agricola, G.; De re metallica; 1556; Engl. translation by H.C. Hoover and L.H. Hoover, Dover Publications, New York, 1950
5. Cooper, D.L., T.P. Cunningham, N.L. Allan, A. McCulloch; Tropospheric lifetimes of potential CFC replacements: rate coefficients for reaction with the hydroxyl radical; *Atm Env*, Vol **26A**, No. 7, pp 1331 - 1334, 1992
6. Prather M. and C.M. Spivakovsky; Tropospheric OH and the lifetime of hydrofluorocarbons; *J Geophys Res*, Vol **95**, No. D11, pp 18723 - 18729, 1990
7. Nimitz J.S. and S.R. Skaggs; Estimating tropospheric lifetimes and ozone-depletion potential of one- and two-carbon hydrofluorocarbons and hydrochlorofluorocarbons; *Environ Sci Technol*, Vol **26**, pp 739 - 744, 1992
8. Hayman, G.D., R.G. Derwent; Atmospheric chemical reactivity and ozone-forming potentials of potential CFC replacements; *Environ Sci Technol*, Vol **31**, pp 327 - 336, 1997.
9. UNEP; The Montreal Protocol on substances that deplete the ozone layer; Final Act (Na. 87-6106), United Nations Environmental Programme, Ozone Secretariate, Nairobi, Kenya, 1987
10. Parker L. and D.E. Gushee; Issue Brief 89021: Stratospheric ozone depletion: regulatory issues; Congressional Research Service, Washington, 1996
11. BOC Gases; Halocarbon 41 (Methyl Fluoride)(CH_3F); U.S. Technical Data Sheets; BOC Ltd., Riverton, 1996
12. Oremland, R.S. and Culbertson, C.W.; Evaluation of methyl fluoride and dimethyl ether as inhibitors of aerobic methane oxidation; *Appl Env Microbiol*, Vol **58**, pp 2983 - 2992, 1992
13. Leisewitz, A. and W. Schwarz; Aktuelle und künftige Emissionen treibhauswirksamer fluorierter Verbindungen in Deutschland; Öko-Recherche GmbH, Forschungsbericht 106 01 074/01, Umweltbundesamt; Berlin, 1996
14. Kroeze, C.; Fluorocarbons and SF_6; Report no. 773001007; Rijksinstituut voor Volksgezondheid en Milieuhygiene (RIVM), Bilthoven (NL), 1995
15. AFEAS/DOE; Energy and global warming impacts of CFC alternative technologies; Alternative Fluorocarbons Environmental Acceptability Study; US Department of Energy, Washington, 1991
16. GREENPEACE; Hydrocarbons and other progressive answers to refrigeration; Greenpeace Germany, Hamburg, 1995
17. Wallington, T.J., W.F. Schneider, D.R. Worsnop, O.J. Nielsen, J. Sehested, W.J. Debruyn, J.A. Shorter; The environmental impact of CFC replacements-HFCs and HCFCs, *Environ Sci Technol*, Vol **28**, pp 320 - 325, 1994
18. STEP-HALOCSIDE/AFEAS; Proceedings of the Workshop "Kinetics and mechanisms for the reaction of halogenated organic compounds in the troposphere"; Dublin, 14 - 16 May 1991, University College Dublin, Dublin, 1991
19. STEP-HALOCSIDE/AFEAS; Proceedings of the Workshop "Kinetics and mechanisms for the reaction of halogenated organic compounds in the troposphere"; Dublin, 23 - 25 March 1993, University College Dublin, Dublin, 1993
20. Ko, M.K.W., N.D. Sze, J.M. Rodriguez, D.K. Weissenstein, C.W. Heisey, R.P. Wayne, P. Biggs, C.E. Canosa-Mas, H.W. Sidebottom, J. Treacy; CF_3 chemistry: potential implications for stratospheric ozone; *Geophys Res Lett*, Vol **21**, pp 101 - 104, 1994
21. Ravishankara, A.R., A.A. Turnipseed, N.R. Jensen, S. Barone, M. Mills, C.J. Howard, S. Solomon; Do hydrofluorocarbons destroy stratospheric ozone?; *Science*, Vol **163**, pp 71 - 75, 1994
22. Pinnock, S., M.D. Hurley, K.P. Shine, T.J. Wallington, T.J. Smyth; Radiative forcing of climate by hydrochlorofluorocarbons and hydrofluorocarbons; *J Geophys Res*, Vol **100**, No. D11, pp 23227 - 23238, 1995
23. Montzka, S.A., R.C. Myers, J.H. Butler, J.W. Elkins, L.T. Lock, A.D. Clarke, A.H. Goldstein; Observations of HFC-134a in the remote troposphere; *Geophys Res Lett*, Vol **23**, No. 2, pp 169 - 172, 1996
24. Oram, D.E., C.E. Reeves, W.T. Sturges, S.A. Penkett, P.J. Fraser, R.L. Langenfelds; Recent tropospheric growth rate and distribution of HFC-134a (CF_3CH_2F); *Geophys Res Lett*, Vol **23**, No. 15, pp 1949 - 1952, 1996

25. McCulloch, A.; Sources of hydrochlorofluorocarbons, hydrofluorcarbons and fluorocarbons and their potential emissions during the next twenty five years; *Environ Monitor Assessm*; Vol **31**, pp 167 - 174, 1994
26. HOECHST; Substitution in sight - HFA 227; Hoechst High Chem Magazine, No. 16, 1995.
27. Gassmann, M; Freon 14 im "Krypton reinst" und in der Atmosphäre; *Naturwissenschaften*, Vol **61**, p.127, 1974
28. Kranz, R.; Organische Fluor-Verbindungen in den Gaseinschlüssen der Wölsendorfer Flußspäte; *Naturwissenschaften*, Vol **53**, pp 593 - 600, 1966
29. Penkett, S.A., N.J.D. Prosser, R.A. Rasmussen, M.A.K. Khalil; Atmospheric measurements of CF_4 and other fluorocarbons containing the CF_3 grouping; *J Geophy Res*, Vol **86**, C6, pp 5172 - 5178, 1981
30. Henry J.L. und R.D. Holliday; Mass spectrometric examination of anode gases from aluminum reduction cells; *J Metals*, pp 1384 - 1385, 1957
31. Rasmussen, R.A., S.A. Penkett, N. Prosser; Measurement of carbon tetrafluoride in the atmosphere, *Nature*, Vol **277**, pp 549 - 551, 1979
32. Fabian, P., R. Borchers, B.C. Krüger, and S. Lal; CF_4 and C_2F_6 in the atmosphere. *J Geophys Res*, Vol 92, pp 9831 - 9835, 1987
33. Cicerone, R.J.; Atmospheric carbon tetrafluoride: a nearly inert gas; *Science*, Vol **206**, pp 59 - 61, 1979
34. Ravishankara, A.R., S. Solomon, A.A. Turnipseed, R.F. Warren; Atmospheric lifetimes of long-lived halogenated species; *Science*, Vol **259**, pp 194 - 199, 1993
35. Wang, W.C., J.P. Pinto, Y.L. Yung; Climatic effects due to halogenated compounds in the Earth's atmosphere; *J Atmos Sci*, Vol **37**, pp 333 - 338, 1980
36. Abrahamson, D.A.; Aluminium and global warming; *Nature*, Vol **356**, p. 484, 1993
37. Cook, E.; Lifetime commitments: Why climate policy-makers can't afford to overlook fully fluorinated compounds; *Issues and Ideas*, World Resource Institute, Washington (USA), 2, 1995
38. Weston, R.E.; Possible greenhouse effects of tetrafluoromethane and carbon dioxide emitted from aluminum production; *Atmos Environ*, **30**, pp 2901 - 2910, 1996.
39. Harnisch, J., R. Borchers, P. Fabian, H.W. Gäggeler, U. Schotterer; Effect of natural tetrafluoromethane; *Nature*, Vol 384, p. 32, 1996
40. Harnisch, J.; Die globalen atmosphärischen Haushalte der Spurengase Tetrafluormethan (CF_4) und Hexafluorethan (C_2F_6); Ph.D.-thesis, University of Göttingen, Cuvillier-Verlag, Göttingen, 1997
41. Woytek, A.J.; Production of CF_4 for electronics industry use; Proceedings of EPA-workshop on PFC (1993), United States Environmental Protection Agency, Washington, 1993
42. Harnisch, J.; Emission scenarios for CF_4 and C_2F_6 derived from reconstructed atmospheric concentrations; Proceedings of Workshop "Successful partnership to reduce PFC emissions from primary aluminum production", United States Environmental Protection Agency, Washington, 1996
43. IPAI; Anode effect and PFC emission survey 1990 - 1993; International Primary Aluminium Institute; London, 1996
44. Beck, T.R.; A non-consumable metal anode for the production of aluminum with low-temperature fluoride melts; *Light Metals*, pp 355 - 360, 1995
45. Penkett, S.A., N.J.D. Prosser, R.A. Rasmussen, M.A.K. Khalil; Atmospheric measurements of CF_4 and other fluorocarbons containing the CF_3 grouping; *J Geophys Res*, Vol **86**, C6, pp 5172 - 5178, 1981
46. Mocella, M.T.; Production and uses of C_2F_6; Proceedings of EPA-workshop on PFC (1993), United States Environmental Protection Agency, Washington, 1993
47. BOC Gases; Halocarbon 218 (Perfluoropropane) (C_3F_8); U.S. Technical Data Sheets; BOC Ltd., Riverton, 1996
48. BOC Gases; Halocarbon 318 (C_4F_8); U.S. Technical Data Sheets; BOC Ltd., Riverton, 1996.
49. Sturges, W., S.A. Penkett, J.M. Lee, K.E. Sturges; In: Stratospheric distribution and lifetimes of halogenated hydrocarbons; Final report to the European Commission, Contract EV5V-CT92-0078, University of East Anglia, Norwich, 1995

50. Morisaki, S.; Simultanous thermogravimetry-mass spectroscopy and pyrolysis-gas chromatography of fluorcarbon polymers; *Thermochimica Acta*, Vol **25**, pp 171 - 183, 1978
51. APME; What is PTFE?; Association of Plastics Manufacturers in Europe, Brussels, 1995
52. Dietz, R.N.; Perfluorocarbon tracer technology; In: Regional and Long-Range Transport of Air Pollution (edited by S. Sandroni), pp 215-247, Elsevier, Amsterdam, 1986
53. Heffter, J.L., T. Yamada, R.N. Dietz; Long-range (CAPTEX) and complex terrain (ASCOT) perfluorocarbon tracer studies; In: Regional and Long-Range Transport of Air Pollution (edited by S. Sandroni), pp 437 - 465, Elsevier, Amsterdam, 1986
54. Piringer, M., K. Baumann, H. Rötzer, J. Riesing, K. Nodop; Results on perfluorocarbon background concentrations in Austria; *Atmos Environ*, Vol **31**, No. **4**, pp 515 - 527, 1997.
55. Harper, D.B. and D.O. O'Hagan; The fluorinated natural products; *Natural Products Report,* Vol **11**, pp 123 - 133, 1994
56. Hall, R.J.; The distribution of organic fluorine in some toxic tropical plants; *New Phytol*, Vol **71**, pp 855 - 871, 1972
57. Frank, H., A. Klein, D. Renschen; Environmental Trifluoroacetate; *Nature*, Vol **382**, p. 34, 1996
58. Zehavi, D. and J.N. Seiber; An analytical method for trifluoacetic acid in water and air samples using headspace gas chromatographic determination of the methyl ester; *Analyt Chem*, Vol **68**, No. 19, pp 3450 - 3459, 1996
59. Teichmann, R.; Fluorhaltige Textilhilfsmittel und der AOX-Wert; Textilveredelung, Vol **31**, 11/12, pp 251 - 254, 1996.
60. Tromp, T.K., M.K.W. Ko, J.M. Rodriguez, N.D. Sze; Potential accumulation of a CFC-replacement degradation product in seasonal wetlands, *Nature*, Vol **376**, pp 327 - 330, 1995
61. Visscher, P.T., C.W. Culbertson, R.S. Oremland; Degradation of trifluoroacetate in oxic and anoxic sediments; *Nature*, Vol **369**, pp 729 - 731, 1994
62. Emptage, M., J. Tabinowski, J.M. Odom; Effect of fluoroacetates on methanogenesis in samples from selected methanogenic environments; *Environ Sci Technol*, Vol **31**, pp 732 - 734, 1997
63. AFEAS; Workshop on the environmental fate of trifluoroacetic acid; Proceedings of the Workshop 3 - 4 March 1994, Miami Beach, Alternative Fluorocarbons Environmental Acceptability Study (AFEAS), Washington, 1994
64. Kanakidou, M., F.J. Dentener, P.J. Crutzen; A global three dimensional study of the fate of HCFCs and HFC-134a in the troposphere; *J Geophys Res*, Vol **100**, D9, pp 18781 - 18801, 1995
65. Wilson, S.R., G. Schuster, G. Helas; Measurements of COFCl and $COCl_2$ near the tropopause; in Ozone in the Atmosphere, edited by R.D. Bojkov und P. Fabian, pp 302 - 305, A. Deepak, Hampton, Va., 1989
66. Zander, R., C.P. Rinsland, E. Mahieu, M.R. Gunson, C.B. Farmer, M.C. Abrams, M.K.W. Ko; Increase of carbonyl fluoride (COF_2) in the stratosphere and its contribution to the 1992 budget of inorganic fluorine in the upper stratosphere; *J Geophys Res*, Vol **99**, D8, pp 16737 - 16743, 1994
67. Huey, L.G., D.R. Hanson, Lovejoy, E.R.; Atmospheric fate of CF_3OH 1: Gas phase thermal decomposition; *J Geophys Res*, Vol **100**, D9, pp 18771 - 18774, 1995
68. Lovejoy, E.R., L.G. Huey, D.R. Hanson; Atmospheric fate of CF_3OH 2: Heterogenous reaction; *J Geophys Res*, Vol **100**, D9, pp 18775 - 18780, 1995
69. Stoiber, R.E., D.C. Leggett, T.F. Jenkins, R.P. Murrmann, W.I. Rose; Organic compounds in volcanic gas from Santiaguito vulcano, Guatemala; *Geol Soc Am Bull*, Vol **82**, pp 2299 - 2302, 1971
70. Private communication DOW-ELANCO.
71. Taylor, A.W.; Post-application volatilization of pesticides under field conditions; *J Air Poll Contr Assoc*, Vol **28**, pp 922 - 927, 1978
72. Francioso, O., E. Bak, N. Rossi, P. Sequi; Sorption of atrazine and trifluralin in relation to the physico-chemical characteristics of selected soils; *Sci Total Env*, 123/124, pp 503 - 512, 1992.
73. Woodrow, J.E., D.G. Crosby, J.N. Seiber; Vapor-phase photochemistry of pesticides; *Res Rev*; Vol **85**, pp 111 - 125, 1983

74. Leitis, E. and D.C. Crosby; Photodecomposition of trifluralin; J Agr Food Chem, Vol 22, No. 5, pp 842 - 848, 1974
75. Braunschweiler, H.; The fate of some pesticides in Finnish cultivated soils; Agric Sci Finl, Vol **1**, pp 37 - 54, 1992
76. Malterre, F., G. Grebil, J.-G. Pierre, M. Schiavon; Trifluralin behaviour in soil: a microlysimeter study; *Chemosphere*, Vol **34**, pp 447 - 454, 1997
77. Golab, T., W.A. Althaus, H.L. Wooten; Fate of [^{14}C] Trifluralin in soil, J Agric Food Chem, Vol **27**, No. 1, pp 163 - 179, 1979
78. Weber, R. and H. Hagenmaier; Synthesis and analysis of mixed chlorinated-fluorinated dibenzo-p-dioxins and dibenzofurans and assessment of formation and occurence of the fluorinated and chlorinated-fluorinated dibenzo-p-dioxins and dibenzofurans; *Chemosphere*, Vol **34**, No. 1, pp 13 - 28, 1997
79. Ko, K.W., N.D. Sze, W.C. Wang, G. Shia, A. Goldman, F.J. Murcray, D.G. Murcray, C.P. Rinsland; Atmospheric sulfur hexafluoride: sources, sinks and greenhouse warming; *J Geophys Res*, Vol **98**, No. D6, pp 10499 - 10507, 1993
80. Lovelock, J.E.; Atmospheric fluorine compounds as indicators of air movements; *Nature*, Vol **230**, p. 379, 1971
81. Krey, P.W., R.J. Lagomarsino, L.E. Toonkel; Gaseous halogens in the atmosphere in 1975; *J Geophys Res*, Vol **32**, No. 12, pp 1753 - 1766, 1977
82. Maiss, M. and I. Levin; Global increase of SF_6 observed in the atmosphere; *Geophys Res Lett*, Vol **21**, pp 569 - 572, 1994
83. Maiss, M., L.P. Steele, R.J. Francey, P.J. Fraser, R.L. Langenfelds, N.B.A. Trivett, I. Levin; Sulfur hexafluoride - a powerful new atmospheric tracer; *Atm Environ*, Vol **30**, pp 1621 - 1629, 1996
84. Geller, L.S., J.W. Elkins, J.M. Lobert, A.D. Clarke, D.F. Hurst, J.H. Butler, R.C. Myers; Tropospheric SF_6: Observed latitudinal distribution and trends, derived emissions and interhemispheric exchange time; Geophys Res Lett, Vol **24**, No. 6, pp 675 - 678, 1997
85. Maiss, M., J. Ilmberger, K.O. Münnich; Vertical mixing in Überlingersee (Lake Constance) traced by SF_6 and heat; *Aquatic Sci*, Vol **56/4**, pp 329 - 347, 1994
86. Maiss, M., J. Ilmberger, A. Zenger, K.O. Münnich; A SF_6 tracer study of horizontal mixing in Lake Constance; *Aquatic Sci*, Vol **56/4**, pp 307 - 328, 1994
87. Wanninkhof, R., J.R. Ledwell, A.J. Watson; Analysis of sulfur hexafluoride in seawater; *J Geophys Res*, Vol **96**, pp 8733 - 8740, 1991
88. Schwander, J., J.M. Barnola, C. Andrié, M. Leuenberger, A. Ludin, D. Raynaud, and B. Stauffer; The age of the air in the firn and the ice at Summit, Greenland; *J Geophys Res*, Vol **98**, D2, pp 2831 - 2838, 1993
89. Harnisch, J., R. Borchers, P. Fabian, M. Maiss; Tropospheric trends for CF_4 and C_2F_6 since 1982 derived from SF_6 dated stratospheric air, *Geophys Res Lett*, Vol **23**, pp 1099 - 1102, 1996
90. BOC Gases; Nitrogen Trifluoride (NF_3); U.S. Technical Data Sheets; BOC Ltd., Riverton, 1996
91. Molina, L.T., P.J. Woodbridge, M.J. Molina; Atmospheric reactions and ultraviolet and infrared absorptivities of nitrogen trifluoride; *Geophys Res Lett*, Vol **22**, No. 14, pp 1873 - 1876, 1995
92. Smith. F.A. and H.C. Hodge; Airborne fluorides and man: part 1; CRC Crit Rev Env Contr, Vol **8**, No. 4, pp 293 - 371, 1979
93. Graedel, T.E. and W.C. Keene; Tropospheric budget of reactive chlorine; *Global Biogeochem Cycles*, Vol **9**, No. 1, pp 47 - 77, 1995
94. Langner, J. and H. Rohde; A global three-dimensional model of the tropospheric sulfur cycle; *J Atm Chem*, Vol **13**, pp 225 - 263, 1991
95. Saether, O.M., B.T. Andreassen, A. Semb; Amounts and sources of fluoride in precipitation over southern Norway; *Atm Env*, Vol **29**, No. 15, pp 1785 - 1793, 1995
96. Mayrhofer, J.; *Z Pflanzenkrankh*, p. 50, Vol **3**, 1893
97. United States Environmental Protection Agency; Air chief - compact disc with emission factors for pollutants; Version 4.0, Washington, 1995

98. Cadle, R.D.; A comparison of volcanic with other fluxes of atmospheric trace gas constituents; *Rev Geophys Space Sci*, Vol **18 (4)**, pp 746-752, 1980
99. Symonds, R.B., W.I. Rose, M.H. Reed; Contribution of Cl- and F-bearing gases to the atmosphere by volcanoes; *Nature*, Vol **334**, pp 415-418, 1987
100. Cicerone, R.J., Halogens in the atmosphere, *Rev Geophys Space Sci*, Vol 19, No. 1, pp 123-139, 1981
101. Vogt, R., P.J. Crutzen, R. Sander; A mechanism for halogen release from sea-salt aerosol in the remote boundary layer; *Nature*, Vol **383**, 1996
102. Sen, B., G.C. Toon, J.-F. Blavier, J.T. Szeto, E.L. Fleming, C.H. Jackman; Balloon-borne observations of mid-latitude hydrofluoric acid; *Geophys Res Lett*, Vol **22**, No. 7, pp 835-838, 1995
103. Arnold, F. and S. Spreng; Balloon-borne mass-spectrometer measurements of HCl and HF in the winter Arctic stratosphere; *Geophys Res Lett*, Vol **21**, No. 13, pp 1255-1258, 1994.
104. Kaye, J.A., A.R. Douglas, C.H. Jackman, R.S. Stolarski, R. Zander, G. Roland; Two-dimensional model calculation of fluorine-containing reservoir species in the stratosphere; *J Geophys Res*, Vol **96**, D7, pp 12865-12881, 1991
105. Holton, J.R., P.H. Haynes, M.E. McIntyre, A.R. Douglas, R.B. Rood, L. Pfister; Stratosphere-troposphere exchange; *Rev Geophys*; Vol **33**, No. 4, pp 403-439, 1995
106. Francis, P., C. Chaffin, A. Maciejewski, C. Oppenheimer; Remote determination of SiF_4 in volcanic plumes: A new tool for volcanic monitoring; *Geophys Res Lett*, Vol **23**, No. 3, pp 249-252, 1996
107. Symonds, R.B. und M.H. Reed; Calculations of multicomponent chemical equilibria in gas-solid-liquid-systems: calculation methods, thermochemical data, and applications to studies of high temperature volcanic gases with examples from Mount St. Helens, *Am J Sci*, Vol **293**, pp 758-863, 1993
108. Graedel, T.E., D.T. Hawkins, L.D. Claxton; Atmospheric chemical compounds - sources, occurence, and bioassay; Academic Press Inc., Orlando, 1986
109. Matthes, S.; Mineralogie; 4th edition, Springer Verlag, Berlin, 1993
110. Grjotheim, K., H. Kvande, K. Motzfeldt, B.J. Welch; The formation and composition of the fluoride emissions from aluminium cells; Paper presented at the International Symposium on the fluoride problem in the aluminium industry; May 24-26, 1972, Trondheim; 1972
111. Grjotheim, K. und B.J. Welch; Aluminium smelter technology; 2nd edition, Aluminium-Verlag, Düsseldorf, 1988
112. Bewers, J.M. and H.H. Hayson; The terrigenous dust contribution to fluoride and iodide in atmospheric precipitation; *J Rech Atm*, Vol **13**, pp 689-697, 1974
113. Goudie, A.S.; Dust storms in space and time; *Prog Phys Geog*, Vol **7**, pp 502-530, 1983
114. Tegen, I. and I. Fung; Modeling of mineral dust in the atmosphere: sources, transport and optical thickness; *J Geophys Res*, Vol **99**, D11, pp 22897-22914, 1994
115. Carpenter, R.; Factors controlling the marine geochemistry of fluorine; *Geochim Cosmochim Acta*; Vol **33**, pp 1153-1167, 1969
116. Wilkniss, P.E. and D.J. Bressan; Chemical processes at the air-sea interface: The behavior of fluorine; *J Geophys Res*, Vol **76**, pp 736-741, 1971
117. Wilkniss, P.E. and D.J. Bressan; Fractionation of the elements of F, Cl, Na, and K at the sea-air interface; *J Geophys Res*, Vol **77**, pp 5307-5315, 1972
118. Barnard and Nordstrom; Fluoride in precipitation-II. Implications for the geochemical cycling of fluorine; *Atm Env*, Vol **16**, No.1, pp 105-111, 1982
119. De Angelis, M. and M. Legrand; Origins and variations of fluoride in Greenland precipitation; *J Geophys Res* , Vol **99**, D1, pp 1157-1172, 1994
120. Gong, S.L., L.A. Barrie, J.M. Prospero, D.L. Savoie, G.P. Ayers, J.P. Blanchet, L. Spacek; Modeling sea-salt aerosols in the atmosphere 2. Atmospheric concentrations and fluxes; J Geophys Res, Vol **102**, D3, pp 3819-3830, 1997
121. Sugiura, T., Y. Mizutani, S. Oana; Fluorine, chlorine and iodine in volcanic gases; *J Earth Sci Nagoya Univ*, Vol **11**, pp 272-278, 1963

122. Mason, B. und C.B. Moore; Principles of Geochemistry; 4. Auflage, John Wiley & Sons, New York, 1982
123. Crutzen, P.J., L.E. Heidt, J.P. Krasnec, W. Pollock, W. Seiler; Biomass burning as a source of atmospheric gases CO, H_2, N_2O, NO, CH_3Cl and COS; *Nature*, Vol **282**, pp 253-256, 1979
124. Bates, T.S., B.K. Lamb, A. Guenther, J. Dignon, R.E. Stoiber; Sulfur emissions to the atmosphere from natural sources, *J Atm Chem*, Vol **14**, pp 315-337, 1992
125. Jambon, A.; Earth degassing and large scale geochemical cycling of volatile elements; in Volatiles in Magmas; M.R. Caroll and J.R. Holloway (eds.); *Rev Mineral*, Vol 30, pp 479-517, 1995
126. IPCC, Climate Change 1995 - The science of climate change; Contribution of working group I to the second assessment report of the Intergovernmental Panel on Climate Change, World Meteorological Organization, Cambridge University Press, Cambridge, 1996
127. Ravishankara, A.R., S. Solomon, A.A. Turnipseed, R.F. Warren; Atmospheric lifetimes of long-lived halogenated species; *Science*, Vol **259**, pp 194-199, 1993
128. HOECHST; HFA 227 pharmaceutical grade; Data Sheet, Hoechst AG, Frankfurt, 1994
129. Roehl, C.M., D. Boglu, C. Brühl, G.K. Moortgat; Infrared band intensities and global warming potentials of CF_4, C_2F_6, C_3F_8, C_4F_{10}, C_5F_{12}, and C_6F_{14}; *Geophys Res Lett*, 22, pp 815-818, 1995
130. WMO; Scientific assessment of ozone depletion: 1994; p. 13.10, World Meteorological Organization, Report No. 37, Geneva, 1994
131. SOLVAY; Schwefelhexafluorid; Solvay-Fluor und Derivate GmbH, Hannover, 1994
132. Grant Thornton LLP; Hydrofluorocarbon-134a (HFC-134a) - Annual Production and sales for the years 1990-1994; Accountant's compilation report for the Alternative Fluorocarbons Environmental Acceptability Study (AFEAS); Vienna (USA), 1995
133. Ullmann's Encyclopedia of Industrial Chemistry; Phosphoric acid and phosphates, 5th edition, Vol **A19**, pp 465-503, VCH Verlagsgesellschaft, Weinheim, 1991
134. United States Bureau of Mines; Annual report; 1992
135. Bayer, G. and H.G. Wiedemann; Fluorrohstoffe - Vorkommen, Verwendung und Probleme; *Chemie in unserer Zeit*, Vol **19**, No. 2, pp 33-41, 1985
136. McCulloch, P.M. Midgley, D.A. Fisher; Distribution of emissions of chlorofluorocarbons (CFCs) 11, 12, 113, 114 and 115 among reporting and non-reporting countries in 1986; *Atm Env*, Vol **28**, No. 16, pp 2567-2582, 1994
137. P.M. Midgley and D.A. Fisher; The production and release to the atmosphere of chlorodifluoromethane (HCFC22); *Atm Env*, Vol **27A**, No. 14, pp 2215-2223, 1993
138. Allmann, R. and S. Koritnig; Fluorine; in Handbook of Geochemistry, edited by K.H. Wedepohl, Vol II-1, Springer Verlag, Berlin, 1978

CHAPTER 4

Iodine Compounds in the Atmosphere

R. Vogt

Ford Forschungszentrum Aachen GmbH, Süsterfeldstr. 200, D-52072 Aachen, Germany.
Email: rvogt@ford.com

The current understanding of iodine sources in the atmosphere, the abundance of atmospheric iodine compounds and their chemistry are reviewed. Besides CH_3I, several other organic iodine compounds of biological origin, such as CH_2I_2, CH_2ClI and C_3H_7I have been detected in ocean water. Due to their low solubility the water can become super-saturated with iodocarbons causing a flux from the aqueous to the gas phase. Anthropogenic sources of iodine in the atmosphere include the combustion of fossil fuels and the potential future use of halon replacements, such as CF_3I. Data from measurements of iodine compounds in ocean water and the atmosphere are compiled. The atmospheric reaction cycles of iodine are described which comprise the photolytic decomposition of iodocarbons, the interchange of iodine between the reservoir species and the interaction of iodine with aerosol particles. Depending on the concentration of reactive iodine in the troposphere, or the lower stratosphere, the photooxidant budget can significantly be affected and substantial ozone depletion is possible.

Keywords: Iodine chemistry, atmosphere, reaction, iodine source, photochemistry, ozone depletion, aerosol.

The Handbook of Environmental Chemistry Vol. 4 Part E
Reactive Halogen Compounds in the Atmosphere
(ed. by P. Fabian and O.N. Singh)

1 Introduction

The atmospheric impact of iodine compounds has received much less attention than the other halogens. Unlike fluorine, chlorine and bromine, the lifetime of organic iodine compounds is in the order of days, or less. For example, methyl iodide, CH_3I, has a photolytic lifetime of only 2–6 days [1]. In contrast to the other halogens, iodine atoms formed upon photolysis do not abstract hydrogen atoms from alkanes because of the low energy content of the H-I bond [2]. The major reaction pathway of I atoms is their reaction with ozone, forming IO radicals. Due to the lack of efficient scavengers for I or IO radicals the chain length of the atmospheric reaction cycles can become long as compared to F, Cl or Br. Already very low levels of iodine might have an impact on the photochemical budgets of the atmosphere.

Generally, it has been assumed that the short-lived organic iodine compounds are not transported to the stratosphere and that the role of iodine is confined to the troposphere [1–4]. However, Solomon and her colleagues [5] speculated on the role of iodine in the depletion of lower stratospheric ozone. It has been suggested that deep convective clouds could transport water-insoluble material from low altitudes to the upper troposphere or lower stratosphere within a time scale of hours [6–8]. Especially in the tropics the photolytically short-lived iodocarbons could, in combination with anthropogenic chlorine and bromine, significantly contribute to the depletion of lower stratospheric ozone [5]. Therefore, the atmospheric impact of free radical iodine chemistry might not be limited to the marine boundary layer or the troposphere, but it also may play a role in the stratosphere.

2 Sources of Iodine Compounds in the Atmosphere

2.1 Oceanic Sources of Iodine

Since the discovery of CH_3I in the marine atmosphere by Lovelock et al. [9] the ocean has been identified as the principle source of atmospheric iodine. Sea water I^- and IO_3^- concentrations are in the order of 0.7×10^{-7} mol l^{-1} or 3×10^{-7} mol l^{-1}, respectively [10–11]. Although iodine is about 10^6 times less abundant in ocean water than chlorine, various pathways of transfer into the atmosphere have been defined.

2.1.1 *Emission of Organic Iodine Compounds from the Ocean*

Several iodocarbons (see Table 1) have been detected in ocean water. Typical I concentrations are in the order of 10^{-9} g l^{-1}. Although the precise mechanism is

Table 1. Surface sea water concentrations of organic iodine compounds at different geographic locations

Compound	Location	Mean Concentration (ng l^{-1})	Ref.
CH_3I	Atlantic Ocean, 35°N–55°S	0.8	[9]
	Eastern Pacific Ocean	1.5	[12]
	Pacific Ocean, Oregon 42–45°N	0.5	[13]
	Atlantic Ocean, 30°S–45°N	0.6	[14]
	Arctic Ocean, Spitzbergen 79°N	0.3–1.0	[15]
	Antarctic Peninsula 60°–67°S	2.6	[16]
CH_2ClI	Antarctic Peninsula 60°–62°S	0.04–0.24	[16]
	Meltwater ponds Mc Murdo, 78°S	0.7–3.4	[17]
	Arctic Ocean Spitzbergen 79°N	0.3	[15]
	N.W. Atlantic ocean 35°–60°N	0.1–3.8	[18]
	Madeira/Gibraltar N. Atlantic 21°N	0.2	[19]
	W. African Coast, 25°N, Maldive Islands 5°N	1	[19]
	Swedish Coastal waters 56°–58°N	0.7–7	[20]
CH_2I_2	Meltwater ponds Mc Murdo, 78°S	5–20	[17]
	Arctic Ocean Spitzbergen 79°N	1.7–6.2	[15]
	Swedisch Coastal waters	2.8–48	[20]
CH_3CHICH_3	Arctic Ocean Spitzbergen 79°N	0.4–1.5	[15]
	Antarctic Ocean, 55°S	0.1	[20]

unknown, it appears that various types of macroalgae and phytoplankton, which may be found in both the coastal water and the open ocean, are able to take up iodide or iodate from the water and transform it into different organic iodine compounds. There is little known about the function of these compounds in macroalgae. It is possible that they are produced because of their antimicrobial effectiveness [21–22], or that their function is to control the growth of epiphytic microorganisms. As a consequence the water becomes locally super-saturated with iodocarbons causing a flux from the aqueous to the gas phase [12].

The best investigated atmospheric iodine compound is methyl iodide. For example, in the south polar sea, especially during springtime, high sea water concentrations were found as compared to other regions [16]. These elevated levels were attributed to the high bioactivity in the Antarctic ocean during springtime. The same conclusion was drawn by Singh et al. [12] who analyzed relatively high CH_3I sea water concentrations at the Peruvian coast, a region which is known for upwelling of cold waters by the Humboldt stream with high bioactivity. On a global basis phytoplankton has been recognized as the dominant source of methyl iodide which has an average sea water concentration in the order of 0.6–0.8 ng l^{-1} [9, 14]. The global oceanic iodine source strength has been estimated to about 0.3–2 Tg $year^{-1}$ [1, 23 57].

There is some evidence for a photochemical production mechanism of CH_3I in sea water as an alternative to biological production. Moore and Zafiriou [57] concluded from laboratory experiments that the recombination of photochemically formed CH_3 an I radicals would constitute a CH_3I source which contributes significantly to the global iodine flux from the ocean. Support for the photochemical production of CH_3I is given by Happell and Wallace [58] who concluded that biological production is not the major factor controlling CH_3I surface water saturation.

Besides methyl iodide, a number of other iodocarbons, such as diiodomethane, chloroiodomethane, or iodopropane have been detected in sea water samples (see Table 1). Especially in coastal waters, high concentrations of CH_2I_2 and CH_2ClI have been reported [19–20]. Class and Ballschmiter [24] showed that, besides several other halogenated methanes, CH_2I_2 is released from brown macro algae (*Fucales sargassum*). In laboratory experiments it has been demonstrated that light induced halogen exchange from CH_2I_2 by sea water Cl^- is possible [19]. In this photochemical reaction CH_2I_2 is converted to CH_2ClI which has also been analyzed in many ocean water samples. Klick and Abrahamsson [20] concluded from their measurements that CH_2I_2 and CH_2ClI are the dominating volatile iodated hydrocarbons in coastal waters. However, Moore and Tokarczyk [18] also detected elevated levels of CH_2ClI in the open sea of the North Atlantic. Their findings indicate that CH_2ClI is a by-product of phytoplankton blooms, and that it is either directly, or indirectly (i.e., via an intermediate) produced. CH_2ClI could be released in a globally significant amount which could be equal to the flux of CH_3I [18].

2.1.2
Iodine Volatilization from the Ocean Surface Layer

A much less determined source of atmospheric iodine may arise from the oxidation of iodide in sea water, producing volatile species such as I_2 or HOI [25–28]. In laboratory experiments, Miyake and Tsunogai [25] irradiated iodide solutions and found that free iodine escapes to the gas phase according to Eq. (1):

$$2\,I^- + 0.5\,O_2 + H_2O + h\nu \rightarrow I_2 + 2\,OH^- \qquad (1)$$

A rough estimation revealed that the global amount of iodine released photochemically from the ocean surface could be in the order of 0.4 Tg $year^{-1}$ [25].

On the basis of laboratory investigations, Garland and co-workers [26–27] showed that sea water iodide is oxidized by ozone, and that it is transferred to the gas phase as molecular I_2. Model calculations suggested that dry deposition of ozone at the ocean surface could represent a quite significant source of iodine which could be in the order of 0.06–0.12 Tg $year^{-1}$ if extrapolated to a global scale. Thompson and Zafiriou [28] pointed out that, according to Eq. (2), HOI is the more likely product of ozone deposition:

$$O_3 + I^- \rightarrow OI^- + O_2 ; OI^- + H^+ \leftrightarrow HOI \quad (2)$$

However, it is difficult to assess quantitatively the HOI, or I_2 flux into the atmosphere because, in these model calculations, iodine diffusion processes and also turbulent mixing within the aqueous surface layer have to be taken into account.

2.2 Anthropogenic Sources of Iodine in the Atmosphere

Anthropogenic activity may also be a source of atmospheric iodine. For example, a significant amount of iodine is released to the atmosphere from the combustion of fossil fuels. The iodine content of coal and petroleum is in the order of 1–4 ppm (i.e., 1–4 mg kg^{-1}) [29]. However, the total source of iodine emitted from the combustion of fossil fuels is less than 2×10^{-2} Tg $year^{-1}$ [1] which is small on a global scale.

Very recently, under the pressure of current international agreement to ban the most ozone-depleting compounds, CF_3I has been discussed as one of the most promising halon replacements which could find use in fire extinguishers. Like other iodocarbons (see Sect. 4) the photolytic lifetime of CF_3I is about two days [30] and it is not expected to penetrate to the stratosphere. Solomon et al. [30] concluded that, even if CF_3I were to be applied as a widespread halon replacement, the amount of iodine released to the atmosphere would be at most 3–7% of the global natural sources of iodocarbons.

3 Atmospheric Observations of Iodine Compounds

The first atmospheric observations of iodine by Heymann, von Fellerberg and Cauer (reviewed by Duce et al. [31]) date back to the 1930s. In these early studies, presumably the sum of particulate and gaseous iodine was measured. Interestingly, during this period most of the atmospheric iodine over Europe originated from the burning of seaweed. In the same studies it was first discovered that the I/Cl ratio of atmospheric samples is up to two or three orders of magnitude higher than the corresponding ratio of sea water which is indicative of a significant difference in the atmospheric cycling of iodine as compared to chlorine. During the past 50 years various analytical techniques, including wet chemical analysis, Neutron Activation Analysis, or Gas Chromatography and Mass Spectrometry have been applied and much more has been discovered about the speciation of the different iodine compounds in the atmosphere.

3.1
Gaseous Organic Iodine

Methyl iodide was the first organic iodine compound to be detected in the atmosphere. During their ship voyage from England to Antarctica, Lovelock et al. [9] discovered CH_3I in the atmosphere and determined a mean mixing ratio of 1.2 pmol mol^{-1}. Globally, a background CH_3I mixing ratio of 1–3 pmol mol^{-1} has been confirmed in several investigations [4, 12, 13, 15, 16] (see Table 2). However, locally very high concentrations of methyl iodide have been reported. For example, Rasmussen et al. [13] measured mixing ratios of 7–22 pmol mol^{-1} in oceanic regions which were attributed to high biomass productivity. Carlier et al. [33] recorded CH_3I levels of up to 43 pmol mol^{-1} at a site close to a coastal macroalgal bed on the northwest coast of France. The highest values were associated with low wind speed conditions, when CH_3I could accumulate. In another study [32], episodes of elevated CH_3I mixing ratios of up to 29–43 pmol mol^{-1} were reported. These periods persisted for a few hours to several days and provide evidence for long distance atmospheric transport of methyl iodide and the probable existence of large source areas, possibly phytoplankton blooms, in the ocean waters.

Table 2. Average mixing ratios of iodocarbons measured in ambient air at different locations

Compound	Location	Average Mixing Ratio (pmol mol^{-1})	Ref.
CH_3I	globally	1–3	[12–13]
	Point Arena, 39°N 125°W	4	[12]
	Iceland coast 66°N	22	[13]
	South African coast, 34°S	12	[13]
	Peru (upwelling), 15°N	7.0	[13]
	Atlantic Ocean, 50°N–60°S	1.2	[9]
	North Sea, 53°N, 1°E	29–43 (max. values)	[32]
	Western tropical N. Pacific, 0°–18°N,	0.5	[4]
	North Pacific Rim, 18°–42°N	1.0	[4]
	Antarctic Peninsula, 60°–66°S, 54°–69°W	2.4	[16]
	Arctic, Spitzbergen, 79°N, 12°E	1.0	[15]
	French Atlantic coast, 53°N	2.6–43	[33]
CH_2I_2	North coast of Teneriffe, 28°N, 17°W	detected	[24]
	Arctic, Spitzbergen, 79°N, 12°E	0.5	[15]
CH_2ClI	North coast of Teneriffe, 28°N, 17°W	detected	[24]
	Arctic, Spitzbergen, 79°N, 12°E	0.1	[15]
	Alert, 83°N, 62°W	0.04 (max. values)	[59]
CH_3CHICH_3	Arctic, Spitzbergen, 79°N, 12°E	2.0	[15]
$CH_3CH_2CH_2I$	Arctic, Spitzbergen, 79°N, 12°E	0.2	[15]
C_2H_5I	East and southeast Asian Sea 35°–15°N, 140°–85°E	0.1	[60]

In order to determine the sum of organic iodine compounds, the ambient air is passed through a particulate filter to strip off particles, and subsequently through a LiOH impregnated filter to absorb inorganic gases, including iodine compounds. Iodocarbons are adsorbed on activated charcoal and analyzed afterwards in the laboratory by Neutron Activation. Total organic iodine concentrations range from 1 to 6 pmol mol^{-1} [34–36] (see Table 3), indicating that CH_3I is the major and often dominating organic iodine compound.

However, recently several other organic iodine compounds, CH_2ClI, CH_2I_2, C_2H_5I, CH_3CHICH_3 and n-C_3H_7I have been detected in the atmosphere (Table 2). Class and Ballschmiter [24] reported the occurrence of CH_2ClI and CH_2I_2 (besides several other organo-halogen compounds) in marine air samples with direct coastal input. Recently, Schall and Heumann [15] quantitatively analyzed organo-iodine compounds in arctic air. On Spitzbergen, Norway, the authors found concentrations in the range 0.004–0.2 pmol mol^{-1}, and 0.1–1.0 pmol mol^{-1} of CH_2ClI or CH_2I_2, respectively. In the same study, i-propyl iodide was the most abundant iodocarbon with concentrations ranging from

Table 3. Average mixing ratios of iodine compounds measured in ambient air at different locations

Compound	Location	Average Mixing Ratio (pmol mol^{-1})	Ref.
Gaseous I	Hawaii, 21°N, 158°W	1.6	[37]
Organic I	Ibaraki, Japan, 37°N, 140°O	1.6–4.1	[35]
	Bermuda, 33°N, 65°W	5.6	[34]
	North Sea, Netherlands	1	[36]
	Antarctica, 20 km inland, 77°S, 50°W	0.1	[36]
Inorganic I	Ibaraki, Japan, 37°N, 140°O	0.2–0.65	[35]
	Bermuda, 33°N, 65°W	3.3	[34]
	North West Territories, Canada	0.1	[34]
	African coast, 0°N, 0°W	2.5	[38]
	Hawaii, 20°N, 155°W	1.0–3.5	[31]
Particle I	Ibaraki, Japan, 37°N, 140°O	0.06–0.7	[35]
	Bermuda, 33°N, 65°W	0.7	[34]
	Hawaii, 21°N, 158° W	0.25–0.5 × gaseous I	[37]
	Hawaii, 20°N, 155°W	0.3–5	[31]
	African coast, 0°N, 0°W	0.8	[38]
	Central Alaska, 66°N, 147°W	0.08–0.16	[39]
	Canadian Arctic, Alert	0.1–0.2	[40]
	North Sea, Netherlands	0.5	[36]
	Antarctica, 20 km inland, 77°S, 50°W	0.1	[36]
	Regensburg, Germany, 49°N, 12°E	0.4–1.5	[41–42]
HI/I_2	North Sea, Netherlands	0.3	[36]
	Antarctica, 20 km inland, 77°S, 50°W	0.1	[36]
HOI	North Sea, Netherlands	0.4	[36]
	Antarctica, 20 km inland, 77°S, 50°W	0.1	[36]
IO	Spitzbergen, spring 95/96, 79°N,12°E	≤ 8	[43]
	Brittany, France, 48°N, 4°E	< 0.4	[44]

0 to 6 pmol mol^{-1}, and an average value of 2 pmol mol^{-1}. At the same time the CH_3I level was about 1 pmol mol^{-1} [15]. This is a good example that under certain circumstances methyl iodide may not be the major carrier of atmospheric iodine.

3.2 Particulate Iodine

Iodine in aerosol particles has been studied at a variety of geographic locations. In most of the marine studies the iodine content of the aerosol particles was in the range 0.1–0.8 pmol mol^{-1}, with the exception of two Hawaiian samples which had a much higher iodine content of 1.8 and 4.8 pmol mol^{-1} [31] (Table 3). Common to all of the studies is the very large enrichment of the I/Cl ratio in marine aerosol which is typically 100–1000 times the sea water ratio [31, 37, 39, 45]. The highest enrichment is seen in the smallest particles [37] and sometimes reaches values as large as 10^4. Although these findings date back to the 1930s, the reason for the enrichment is still unclear. Most likely at least two mechanisms operate, one causing an initial enrichment when larger aerosol particles are first generated at the ocean surface involving an iodine-rich surface film. By the second mechanism the longer lived small particles would scavenge organic and inorganic iodine compounds from the gas phase [23] which would result in an iodine enrichment, especially in the smaller particle size range.

Commonly, Neutron Activation Analysis is applied as analytical method. With this very sensitive technique, however, it is not possible to distinguish between the different chemical forms of iodine. Two oxidation states, I^- and IO_3^-, are accessible in the deliquesced atmospheric aerosol. Apart from thermodynamic calculations [37], there is a recent study by Wimschneider and Heumann [42] who applied Isotope Dilution Mass Spectrometry in combination with Anion Exchange Chromatography to determine iodide and iodate concentrations of aerosol particles. In the size-fractionated marine aerosol samples which were collected over the South Atlantic the prevailing iodine species was iodate [42]. A relatively higher iodate content was also reported in aerosol samples collected in Regensburg, Germany, with the exception of the larger size fractions (i.e., particle diameter > 3 µm) which were dominated by iodide. This can be explained by a contribution from iodide of soil dust, while iodate contained in the smaller size fraction would be the result of atmospheric scavenging and chemical conversion within the longer-lived small aerosol particles [42].

3.3 Gaseous Inorganic Iodine Compounds

In the first studies of gaseous inorganic iodine elemental analytic techniques were applied [31, 34, 38]. The sampled air is passed through a particulate filter to collect particles including particulate iodine. In the second stage the air stream is bubbled through an aqueous K_2CO_3, or KOH solution, or alternatively passed through an impregnated filter to absorb inorganic iodine compounds. Subsequently, the sum of elemental iodine is analyzed by Neutron Activation.

The sum of inorganic iodine is in the range 0.1–4 pmol mol^{-1} with the lowest values found at a remote continental site in the Northwest Territories of Canada [34]. In the marine boundary layer typically much larger concentrations of inorganic iodine have been reported [31, 38]. For example, along the African coast Rancher and Kritz [38, 45] found average concentrations of about 2.5 pmol mol^{-1}. In this study about twice as much gaseous inorganic iodine was measured during daylight hours as compared to night. During the same cruise the particulate iodine averaged 0.8 pmol mol^{-1} and did not show such day/night variations. A satisfying explanation for the diurnal variation of inorganic iodine has not yet been found, although several aspects have been investigated in a modeling study [3]. On the other hand, Moyers and Duce [37] did not observe comparable day/night variations of gaseous iodine in the Hawaiian atmosphere.

In a recent study Gäbler and Heumann [36] used a specifically prepared filter system in connection with Isotope Dilution Mass Spectrometry to differentiate among the gaseous iodine species HI/I_2 and HOI, and organic iodine as well as particulate iodine. Both HI/I_2 and HOI represented a significant amount of the total gaseous iodine, up to almost 50% in one sample from Antarctica [36].

There have been several attempts to detect iodine oxide radicals, IO, in the marine atmosphere [43, 44]. IO is expected to be an important reaction intermediate in the atmospheric iodine cycle (see Sect. 4) and therefore measurements of IO would be a good test for the applicability of atmospheric iodine models. At a coastal site in western Brittany [43] no IO could be detected and an upper limit of 0.4 pmol mol^{-1} was reported. In zenith sky DOAS measurements [44] which interrogated both the troposphere and the stratosphere over Spitzbergen, the absorption features of IO were found. However, at the present state the data is preliminary, although an upper limit of 8 pmol mol^{-1} was given under the assumption that all IO was present in the polar boundary layer of 1.5 km height.

4 Atmospheric Chemistry of Iodine Compounds

4.1 The Atmospheric Reaction Cycles

Once released to the atmosphere photolytic decomposition, Eq. (3), is the predominant fate of organic iodine compounds, R-I:

$$\text{R-I} + h\nu \rightarrow \text{R} + \text{I} \tag{3}$$

Alternatively, gas phase oxidation by OH radicals is of less importance because of the relatively slow OH reaction rate. Surface photolysis rates, which are calculated from absorption cross sections and the solar irradiance spectrum, are given in Table 4. While the atmospheric lifetime of iodocarbons containing one I atom is in the order of one to five days, photolysis of CH_2ClI or CH_2I_2 occurs at a much faster rate and is in the order of hours to minutes [47].

Unlike the other halogen atoms, which might be scavenged by reaction with various organic species, I atoms do not abstract hydrogen from saturat-

Table 4. Photolysis rate constants and instantaneous atmospheric lifetime of organic iodine compounds at a solar zenith angle of 40°

Compound	Photolysis Rate (s^{-1})	Lifetime	Ref.
CH_3I	5×10^{-6}	55 hours	[46]
$CH_3CH_2CH_2I$	7×10^{-6}	40 hours	[47]
CH_3CHICH_3	2×10^{-5}	14 hours	[47]
CF_3I	2×10^{-5} [a]	14 hours [a)]	[30]
CH_2ClI	1×10^{-4}	3 hours	[47]
CH_2I_2	5×10^{-3}	3 min	[47]

[a] The data for CF_3I is calculated for midlatitudes in winter.

ed hydrocarbons. The predominant fate of I atoms is reaction with ozone (Eq. 4):

$$I + O_3 \rightarrow IO + O_2 \tag{4}$$

The time scale of this reaction is about 2 s at an ozone mixing ratio of 20 nmol mol^{-1} [46]. At low NO_x levels, which are typically observed in the unpolluted marine boundary layer, the major fate of iodine oxide is photodissociation to regenerate I atoms (Eq. 5):

$$IO + h\nu \rightarrow I + O \tag{5}$$

The lifetime of IO with respect to photodissociation at 40 °N is approximately 4 s [48]. The reaction cycle, Eqs. (4) and (5), has no net chemistry, since O atoms regenerate O_3:

$$O + O_2 + M \rightarrow O_3 + M \tag{6}$$

At the higher NO_x levels, which can occur in the coastal marine boundary layer (up to approximately 1 nmol mol^{-1}), the reaction of IO with NO, (Eq. 7), becomes significant. However, following photodissociation of NO_2, and regeneration of O_3, via Eq. (6), there is another reaction cycle closed which has no net chemistry:

$$I + O_3 \rightarrow IO + O_2 \tag{4}$$
$$IO + NO \rightarrow I + NO_2 \tag{7}$$
$$NO_2 + h\nu \rightarrow NO + O \tag{8}$$
$$O + O_2 + M \rightarrow O_3 + M \tag{6}$$

To a first approximation the ratio, IO/I, is determined by the relative rates of the reactions which convert I into IO (Eq. 4), and IO back into I (Eqs. 5 and 7). Because of the low NO_x concentration in the marine atmosphere IO is the predominating radical iodine species. The calculated IO/I ratio is approximately 1–10, depending on NO_x and O_3 levels [3, 4, 46, 49]. Due to the fast photolysis reaction (Eq. 5), this ratio is significantly lower than that of the other halogens, BrO/Br or ClO/Cl, which are in the order of 100 and 1000, respectively [50, 51].

Unlike the two "null cycles" described above, the reactions of IO with HO_2, ClO, BrO, or IO represent reaction cycles with net chemistry which may lead to very efficient catalytic ozone removal. At low IO levels the reaction with HO_2 (Eq. 9) followed by HOI photolysis (Eq. 10), is most important:

$$IO + HO_2 \rightarrow HOI + O_2 \quad (9)$$
$$HOI + h\nu \rightarrow OH + I \quad (10)$$
$$I + O_3 \rightarrow IO + O_2 \quad (4)$$
$$HO_2 + O_3 \rightarrow 2\,O_2 + OH \text{ (cycle I)} \quad (11)$$

At $[HO_2] = 2 \times 10^8$ molecules cm^{-3}, which is a typical concentration in the remote marine boundary layer [51], the rate-limiting step of this cycle is the photolysis of HOI (Eq. 10). Although the absorption cross sections of HOI are not well established, the HOI photolysis rate constant has been estimated in the order of $5-6 \times 10^{-3}\ s^{-1}$ [4, 46].

With increasing iodine oxide levels a cycle involving the IO self reaction becomes important:

$$IO + IO \rightarrow I + I + O_2 \quad (12a)$$
$$IO + IO + M \rightarrow I_2O_2 + M \quad (12b)$$
$$I_2O_2 + h\nu \rightarrow 2\,I + O_2 \quad (13)$$
$$I + O_3 \rightarrow IO + O_2 \quad (4)$$
$$2O_3 \rightarrow 3O_2 \quad \text{(cycle II)}$$

In the atmosphere the contribution of this cycle increases with square of the IO concentration, because the IO self reaction (Eq. 12) is the rate limiting step. Although the product branching in Eq. (12) is not well known it can be assumed that any I_2O_2 formed in Eq. (12b) will rapidly be photolysed in the sunlit atmosphere [52]. For example, at $[IO] = 2 \times 10^7$ molecules cm^{-3} the first order rate constant of the rate determining step (Eq. 12) is $4 \times 10^{-3}\ s^{-1}$ [48, 52], which is comparable to the HOI photolysis rate. At higher IO levels reaction cycle (II) overruns cycle(I).

Two further reaction cycles in which ozone is catalytically destroyed involve the reaction of IO with the halogen oxide radicals, BrO, and ClO [5]:

$$IO + BrO \rightarrow I + Br + O_2 \quad (14)$$
$$I + O_3 \rightarrow IO + O_2 \quad (4)$$
$$Br + O_3 \rightarrow BrO + O_2 \quad (15)$$
$$2O_3 \rightarrow 3O_2 \quad \text{(cycle III)}$$
$$IO + ClO \rightarrow I + Cl + O_2 \quad (16)$$
$$I + O_3 \rightarrow IO + O_2 \quad (4)$$
$$Cl + O_3 \rightarrow ClO + O_2 \quad (17)$$
$$2O_3 \rightarrow 3\,O_2 \quad \text{(cycle IV)}$$

The importance of these cycles has been pointed out by Solomon and coworkers [5, 30]. The authors estimated that in the lower stratosphere (10–20 km) ozone loss is dominated by these two cycles if the rate of the IO + BrO and IO + ClO reactions (Eqs. 14 and 16) would approach the gas collision frequency (1×10^{-10} cm^3 molecule s^{-1}). First experimental kinetic studies [53–54] of reactions at Eqs. (14) and (16) have confirmed that the cross halogen reactions are extremely fast, although it appears that the IO+ClO reaction might have been overestimated in the model calculations by a factor of five [5, 30].

In the reactions of IO + HO_2 (Eq. 9) and IO + IO (Eq. 12b) the temporary iodine reservoir species, HOI or I_2O_2, respectively, are formed. Other reservoir species, HI, $IONO_2$ and INO_2, are products of the radical termination reactions (Eqs. 18–20) [4, 46]:

$$I + HO_2 \rightarrow HI + O_2 \quad (18)$$
$$IO + NO_2 + M \rightarrow IONO_2 + M \quad (19)$$
$$I + NO_2 + M \rightarrow INO_2 + M \quad (20)$$

The relative abundance of the gas phase reservoir species depends on both their formation rate, and on the rate of reactions at Eqs. (21)–(24) which regenerate iodine atoms:

$$OH + HI \rightarrow I + H_2O \quad (21)$$
$$IONO_2 + h\nu \rightarrow I + NO_3 \quad (22)$$
$$IONO_2 + M \rightarrow IO + NO_2 + M \quad (23)$$
$$INO_2 + h\nu \rightarrow I + NO_2 \quad (24)$$

A schematic diagram of the atmospheric iodine cycling is shown in Fig. 1. The reservoir species I_2O_2, HOI, $IONO_2$, INO_2, and HI can also be scavenged by the aerosol and are accumulated as particulate iodide and iodate. The chemical steps involved in these hydrolysis and oxidation reactions are not well known [3–4]. However, scavenging of the inorganic iodine species from the gas phase is a likely explanation for the strong enrichment of iodine observed in marine aerosols (see Sect. 3.2). Depending on the aerosol lifetime, particulate iodine is transferred back to the ocean where it is part of the marine geochemical cycles [55].

4.2 Partitioning of Iodine Species

As discussed in more detail in Sect. 3, organic iodine compounds are the predominant form of atmospheric iodine with a total concentration which is in the order of 1–5 pmol mol^{-1}. Gaseous inorganic iodine and especially iodine in particulates is typically found at somewhat lower levels.

The partitioning among the different inorganic iodine compounds can be calculated in box model simulations. In the more recent modeling studies [3–4, 46], HOI is estimated to be the predominant inorganic species at concentrations of 0.5 pmol mol^{-1} or less. Depending on the model parameters the calculated HI, IO, and I concentrations are in the order of 0.1–0.01 pmol mol^{-1}. Absolute levels of the individual iodine species depend on the total iodine load of the air which is determined by the total flux of iodine from the ocean surface. Jenkin [46] estimated the steady state concentrations of individual iodine species as a func-

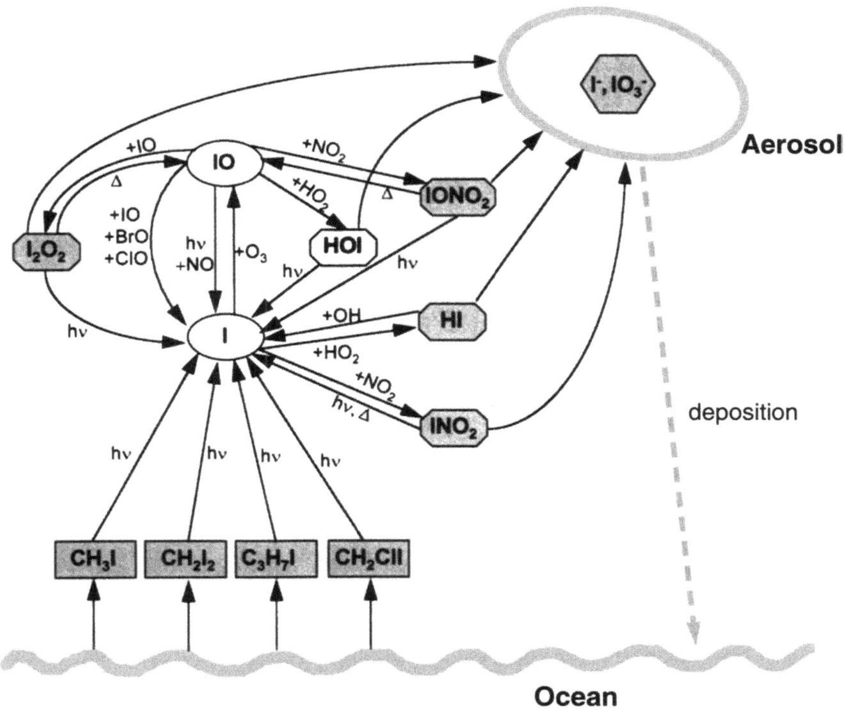

Fig. 1. Tropospheric iodine photochemical reaction scheme. Organic iodine compounds are shown in *rectangular boxes* and iodine reservoir species in *rounded rectangles*. Particulate iodine in the marine aerosol is represented by a *hexagon*

tion of NO_x. Most affected by the NO_x levels is the reservoir species $IONO_2$, of which the concentration increases from 0.01 to 0.2 pmol mol^{-1} if NO_x levels were at 10 or 1000 pmol mol^{-1}, respectively.

In modeling studies it is assumed that particulate iodine is formed as a product of gas phase iodine scavenging [3, 4, 46, 49]. However, due to the lack of laboratory data the parameters needed to model such processes rely heavily on many assumptions or comparison with similar scavenging reactions of other halogen compounds, which makes a quantitative assessment of the particulate iodine abundance difficult.

4.3
Potential Impact of Iodine on the Tropospheric Photochemistry

4.3.1
Modulation of the HO_2/OH Ratio

The reaction cycles described in the last section potentially impact boundary layer levels of the HO_x species (i.e., HO and HO_2), the NO_x species (i.e., NO and NO_2) and O_3. In Cycle (I) for example, HO_2 is converted into OH which would lower the HO_2/OH ratio and therefore impact the atmospheric photooxidant budget significantly. Davis et al. [4] investigated the impact of iodine chemistry on the HO_2/OH ratio in three scenarios with reactive iodine levels, I_x ($I_x = I + IO + HOI + HI + 2\,I_2O_2 + INO_2 + IONO_2$), of 0.5, 1.5 and 7 pmol mol^{-1}, corresponding to marine regions with low productivity, moderate productivity (i.e., including other iodocarbons than CH_3I) and with high productivity, respectively. Only minor changes of the HO_2/OH ratio were found when I_x levels were limited to 0.5 and 1.5 pmol mol^{-1}. However, the HO_2/OH ratio would shift by as much as –18% to –24% if I_x = 7 pmol mol^{-1} was assumed [4].

4.3.2 *Ozone Destruction*

As it has been described in more detail in Sect. 4.1, the reaction cycles (I to IV) result in net ozone depletion. In their recent modeling study Davis et al. [4] estimated reductions of the total tropospheric ozone column between 5 and 30% if I_x levels of 1.5 or 7 pmol mol^{-1}, respectively, were assumed. Ozone destruction would be insignificant if I_x would be in the order of 0.5 pmol mol^{-1}. Jenkin [46] came to a similar conclusion that even at [CH_3I] = 30 pmol mol^{-1} iodine chemistry would only cause about 13% additional ozone destruction. The majority of this iodine related ozone depletion would be caused by the $IO + HO_2$ reaction, Cycle (I).

As in the lower stratosphere, where chlorine and bromine chemistry in combination with iodine (Cycles (II) and (IV)) can lead to significant ozone depletion [5], the role of the cross halogen cycles in the troposphere or marine boundary layer is a subject of discussion. Solomon et al. [5] have speculated on the potential role of iodine chemistry in the annually observed depletion of boundary layer ozone during Arctic spring. Concurrently with the ozone depletion, high levels of BrO radicals were observed [56]. Therefore, the IO + BrO reaction (Cycle (III)) could lead to very efficient ozone depletion. In a recent modeling study of the halogen chemistry during Arctic polar sunrise, Sander et al. [49] concluded that the BrO+IO reaction (Eq. 14) could significantly contribute to ozone depletion. Assuming IO levels of 1 pmol mol^{-1}, it was estimated that the ozone mixing ratio during a depletion event would be about 5 nmol mol^{-1} lower as if iodine were not present. However, for a realistic quantitative assessment of the impact of iodine chemistry on the ozone budget, ambient measurements of the key iodine species, such as IO, HOI, or HI would be needed.

5 Summary

The major iodine source in the atmosphere is the emission of organic iodine compounds from the oceans. Organic iodine compounds are formed in biogenic processes of certain macroalgae and phytoplankton. Because of the supersaturation of the surface water these compounds evaporate and can be detected in the marine air. For example, the global average concentration of methyl iodide is 1–3 pmol mol^{-1}, although elevated levels of up to 43 pmol mol^{-1} have been reported. Other iodocarbons which have been detected in the atmosphere are CH_2I_2, CH_2ClI, C_2H_5I, and C_3H_7I. All these iodine species photodissociate during daytime and have an estimated lifetime in the order of minutes to days. Iodine atoms which are released upon photolysis initiate various atmospheric reaction cycles which may impact photooxidant concentrations, both in the troposphere and stratosphere. The magnitude of ozone depletion, or the shifting of the atmospheric HO_2/OH ratio largely depend on the total amount of reactive iodine present which is generally not well known.

6 References

1. Chameides W L, Davis D D (1980) J Geophys Res 85:7383
2. Jenkin ME, Cox RA, Candeland DE (1985) J Atmos Chem 2:359
3. Chatfield RB, Crutzen PJ (1990) J Geophys Res 95D:22319
4. Davis DD, Crawford J, Liu S, McKeen S, Bandy A, Thornton D, Rowland F, Blake D (1996) J Geophys Res 101:2135
5. Solomon S, Garcia RR, Ravishankara AR (1994) J Geophys Res 99D:20491
6. Gidel LT (1983) J Geophys Res 88:6587
7. Pickering K, Thompson AM, Scala JR, Tao WK, Simpson J (1992) J Atmos Chem 14:297
8. Chatfield RB, Crutzen PJ (1984) J Geophys Res 89:7111
9. Lovelock JE, Maggs RJ, Wade RJ (1973) Nature 241:194
10. Tsunogai, S 1971 Deep Sea Res 18:913
11. Ullman WJ, Luther GW III, De Lange G J, Woittiez JRW (1990) Marine Chemistry 31:153
12. Singh HB, Salas LJ, Stiles RE (1983) J Geophys Res 88:3684
13. Rasmussen RA, Khalil MA, Gunawarda R, Hoydt SD (1982) J Geophys Res (87):3086
14. Tanzer D, Heumann KG (1992) Intern J Environ Anal Chem 48:17
15. Schall C, Heumann KG (1993) Fres J Anal Chem 346:717
16. Reifenhaeuser W, Heumann KG (1992) Atmos Environ 26 A: 2905
17. Schall Ch, Heumann KG, De Mora S, Lee PA (1996) Antarctic Science 8:45
18. Moore RM, Tokarczyk R (1992) Geophys Res Lett 19:1779
19. Class T, Ballschmiter K (1987) Fres J Anal Chem 327:39
20. Klick S, Abrahamson K (1992) J Geophys Res 97:12683
21. Schall C, Laturnus F, Heumann, KG (1994) Chemosphere 28:1315
22. Fenical W (1981) Natural halogenated organics. In: Duutsma E K, Dawson R (eds) Marine organic chemistry. Elsevier New York, p 375
23. Cicerone RJ (1981) Rev of Geophys and Space Phys 19:123
24. Class T, Ballschmiter K (1988) J Atmos Chem 6:35
25. Miyake Y, Tsunogai (1963) J Geophys Res 68: 3989
26. Garland JA, Elzerman AW, Penkett SA (1980) J Geophys Res 85:7488
27. Garland JA, Curtis H (1981) J Geophys Res 86:3183.
28. Thompson AM, Zafiriou OC (1983) J Geophys Res 88 C:6696
29. Valcovic V (1978) Trance Elements in Petroleum, PPC Books, Tulsa
30. Solomon S, Burkholder JB, Ravishankara AR, Garcia RR (1994) J Geophys Res 99:20929
31. Duce RA, Winchester JW, Van Nahl TW (1965) J Geophys Res 70:1775
32. Oram DE, Penkett SA (1994) Atmos Environ 28:1159
33. Carlier P, Fresnet P, Pashalidis S, Tsetsi M, Martinet A, Lescoat V, Dupont B, Chebbi A, Girard R (1991) Air Pollution Research Report 35:167
34. Rahn KA, Borys RD, Duce RA (1976) Science 192: 549
35. Yoshida S, Muramatsu Y (1995) J Radioanalytical and Nucl Chem 196:295
36. Gaebler HE, Heumann, KG (1993) Fres J Anal Chem 345:53
37. Moyers JL, Duce RA (1972) J Geophys Res 77:5229
38. Kritz MA, Rancher J (1980) J Geophys Res 85:1633

39. Sturges WT, Shaw GE (1993) Atmos Environ 27 A: 2969
40. Sturges WT, Barrie LA (1988) Atmos Environ 22:1179
41. Gaebler H-E, Heumann KG (1993) Intern J Environ Anal Chem 50:129
42. Wimschneider A, Heumann KG (1995) Fresenius J Anal Chem 353:191
43. Brauers T, Dorn H-P, Platt U (1990) Physical chemical behavior of atmospheric pollutants. Air Pollution Res Rep 23:237
44. Wittrock F, Eisinger M, Ladstätter-Weißenmaier A, Richter A, Burrows JP (1996) Annales Geophysicae Suppl II 14:C607
45. Rancher J, Kritz MA (1980) J Geophys Res 85 C: 5581
46. Jenkin ME (1992) Tech Rep AEA-EE-0405 United Kingdom Atomic Energy Authority, Harwell Laboratory, Oxon, OX11 0RA, UK
47. Roehl CM, Burkholder JB, Moortgat GK, Ravishankara AR, Crutzen PJ (1997) J Geophys Res 102:12819
48. Laszlo B, Kurylo MJ, Huie RE (1995) J Phys Chem 99:11701
49. Sander R, Vogt R, Harris GW, Crutzen PJ (1997) Tellus 49 B:522
50. Platt U, Janssen C (1995) J Chem Soc Faraday Trans 100:175
51. Vogt R, Crutzen PJ, Sander R (1996) Nature 383:327
52. Harwood MH, Burkholder JB, Hunter M, Fox RW, Ravishankara AR (1997) J Phys Chem 101:853
53. Huie RE, Laszlo B, Kurylo MJ, Buben SN, Trofimova EM, Spassky AI, Messineva NA, Nevozhai D, Miziolek AW (1995) The Atmospheric Chemistry of Iodine Monoxide Halon Options Technical Working Conference, Albuquerque, NM, May 10 1995
54. Laszlo B, Huie RE, Kurylo MJ (1997) J Geophys Res 102:1523
55. Wong GTF (1991) Rev in Aquatic Sci 4:45
56. Barrie LA, Bottenheim JW, Schnell RC, Crutzen PJ, Rasmussen, RA (1988) Nature 334:138
57. Moore RM, Zafiriou OC (1994) J Geophys Res 99:16415
58. Happell JD, Wallace WR (1996) Geophys Res Lett 23:2105
59. Yokouchi Y, Barrie LA, Toom D, Akimoto H (1996) Atoms Environ 30:1723
60. Yokouchi Y, Mukai H, Yamamoto H, Otsuki A, Saitoh C, Nojiri Y (1997) J Geophys Res 102:8805

CHAPTER 5

Properties and Applications of Industrial Halocarbons

P.M. Midgley[1] · A. McCulloch[2]

[1] M & D Consulting, Ludwigstrasse 49, D-70771 Leinfelden, Germany.
E-mail: paulinemidgley@compuserve.com
[2] ICI Chemicals & Polymers Ltd., Safety & Environment Department, P.O. Box 13, The Heath, Runcorn, Cheshire, WA7 4QF, UK. *E-mail: archie_mcculloch@ici.com*

Chlorofluorocarbons (CFCs) were developed in the 1920s and 1930s to replace hydrocarbon and other refrigerants that, by the engineering standards of the time, were considered unsafe. Growth into applications in many market and industrial sectors followed and, with the phaseout of CFCs under the Montreal Protocol, there is a significant market for replacement products such as halocarbon alternatives or the historical fluids – hydrocarbons, ammonia and water. The factors determining the choice of replacement product are discussed. Scenarios for future emissions of halocarbons are also developed. These are based on both the underlying market demand shown by the historical growth of CFCs and HCFCs (hydrochlorofluorocarbons) and on the extents to which that demand is met by each group of substitute materials and by improvements to technology, such as better containment and effective recycling. The atmospheric concentrations of alternative fluorocarbons (particularly hydrofluorocarbons, HFCs) resulting from these scenarios are compared with previous work from the Intergovernmental Panel on Climate Change that relied on macro-economic estimates.

Keywords: Aerosol propellant, Air conditioning, Fire extinguisher, Insulating foam, Refrigeration.

List of Abbreviations and Symbols

AFEAS	Alternative Fluorocarbons Environmental Acceptability Study
AGAGE	Advanced Global Atmospheric Gases Experiment
ALE	Atmospheric Lifetime Experiment

The Handbook of Environmental Chemistry Vol. 4 Part E
Reactive Halogen Compounds in the Atmosphere
(ed. by P. Fabian and O.N. Singh)

CDIAC	Carbon Dioxide Information Analysis Center
CFC	chlorofluorocarbon
DOE	Department of Energy (USA)
EESC	equivalent effective stratospheric chlorine
EU	European Union
GAGE	Global Atmospheric Gases Experiment
GBEC	12th German Bundestag's Enquete Commission on Man and the Environment
GDP	gross domestic product
HCFC	hydrochlorofluorocarbon
HFC	hydrofluorocarbon
Halon	fully saturated compound containing inter alia C, Br and F and used for fire-extinguishing
IPCC	Intergovernmental Panel on Climate Change
OECD	Organisation for Economic Co-operation and Development
PAFT	Program for Alternative Fluorocarbons Toxicology (Testing)
PFC	perfluorocarbon
SORG	Stratospheric Ozone Review Group (UK)
UNEP	United Nations Environment Programme
USEPA	Environment Protection Agency of the USA
WMO	World Meteorological Organisation

1 Introduction and History

Chlorofluorocarbons (CFCs) were introduced in the 1930s to replace refrigerants such as chloromethane (introduced 1878), carbon tetrachloride (1912) and isobutane and propane (1920s) that were considered too dangerous for domestic appliances [1]. The CFCs were the result of a deliberate campaign to identify suitable materials that, from the outset, concentrated on compounds comprising the early members of Groups IV – VII of the periodic table (H, C, N, O, F, S, Cl and Br), compounds of the other elements being insufficiently volatile, unstable, toxic or permanent gases [2]. From the 1930s until the mid-1980s, CFCs were the refrigerants of choice and equipment was perfected to make optimum use of their properties within the limits of contemporary engineering and economics. Replacements, whether provided by other chemical agents or other technologies, need to match the performance of CFCs.

Table 1 lists the CFCs and potential fluorocarbon alternatives, together with one of their most important physical properties – volatility, characterised here by the boiling point at atmospheric pressure. In the table, the past and present commercial importance of the substances has been categorised into three groups: those sufficiently important environmentally and commercially to be part of a data collection activity, those where data collection might have been expected but which do not meet the reporting criteria given in Sect. 2.2 of the chapter on the Production, Sales and Emissions of Industrial Halocarbons (this volume) and those that are not commercially available.

Table 1. Halocarbon properties

	Compound	Chemical formula	Boiling point at 1 atmosphere °C[a]	Atmospheric lifetime (years)	ODP relative to CFC-11[b]	GWP relative to CO_2 at 100 years[c,d]
Compounds for which production and sales data have been collected	CFC-11	CCl_3F	23.8	50	1	4000
	CFC-12	CCl_2F_2	– 29.8	102	1	8500
	CFC-113	CCl_2FCClF_2	47.6	85	0.8	5000
	CFC-114	$CClF_2CClF_2$	3.6	300	1	9300
	CFC-115	$CClF_2CF_3$	– 38.7	1700	0.6	9300
	HCFC-22	$CHClF_2$	– 40.8	13.3	0.055	1700
	HCFC-141b	CH_3CCl_2F	32	9.4	0.11	630
	HCFC-142b	CH_3CClF_2	– 9.2	19.5	0.065	2000
	HCFC-124	$CHClFCF_3$	– 12	5.9	0.022	480
	HFC-134a	CH_2FCF_3	– 26.5	14.6	0	1300
	Methyl chloroform	CH_3CCl_3	74.1	5.4	0.1	110
	Halon-1211	$CBrClF_2$	– 4	20	3	No data
	Halon-1301	$CBrF_3$	– 57.8	65	10	5600
Compounds which do not meet criteria for data collection	FC-218[e]	$CF_3CF_2CF_3$	– 36	2600	0	7000
	HFC-43-10mee	CF_3CHF-$CHFCF_2CF_3$	45	17.1	0	1300
	HFC-227ea[f]	CF_3CHFCF_3	– 16.5	36.5	0	2900
	HFC-152a	CH_3CHF_2	– 24.7	1.5	0	140
	HFC-143a	CH_3CF_3	– 47.6	48.3	0	3800
	HFC-125	CHF_2CF_3	– 48.5	32.6	0	2800
	HFC-32	CH_2F_2	– 51.7	5.6	0	650
	HFC-23	CHF_3	– 82.1	264	0	11700
	HCFC-123	$CHCl_2CF_3$	27.1	1.4	0.022	93
	Carbon tetrachloride	CCl_4	76.8	42	1.1	1400
Compounds not commercially available	HCFC-225ca[g]	$CHCl_2CF_2CF_3$	51.1	2.5	0.025	170
	HCFC-225cb[g]	$CHFClCF_2CF_2Cl$	56.1	6.6	0.033	530
	HCF-236fa[e]	$CF_3CH_2CF_3$	– 0.7	209	0	6300
	HFC-245ca[h]	$CHF_2CF_2CFH_2$	25.4	6.6	0	560

[a] From [3], unless otherwise stated; [b] From [4]; [c] From [5]; [d] From [6]; [e] Boiling point from [7]; [f] Boiling point from [8]; [g] Boiling point from [9]; [h] Boiling point from [7].

The first group comprises most of the substances in the Montreal Protocol and also HFC-134a. Leaving aside the halons for which volatility, although important, was secondary to fire-extinguishing efficiency, it is readily apparent that the original CFCs and methyl chloroform spanned a wide range of volatility, with boiling points ranging from –38.7 °C to +74.1 °C. This enabled applications from deep-freeze refrigeration through to solvents that would vaporise at higher than room temperature. While boiling point and volatility are not the sole determinants of suitability, they do serve as good indicators of what is not suitable. Thus, while there are a number of possible hydrofluorocarbon (HFC) replacements for CFC-12, the most ubiquitous refrigerant, many of the potential HFCs of the second group are too volatile for the freezing and chilling applications for which CFC-12 is technically ideal. The simplest result of using too volatile a refrigerant is that the system pressure is increased, leading to higher equipment costs and possibly higher operating costs. Several of the alternative HFCs can be used in low temperature refrigeration but the need for this is limited and so the quantities involved are comparatively small, and likely to remain so. Replacement of CFC-11, CFC-113 and methyl chloroform is even more problematic because of the wide differences between their volatilities and those of the available new compounds and it is possible that many of the effects that these materials provide will be substituted by new technologies.

Tetrafluoroethane, HFC-134a, is the sole HFC that has sufficiently large sales to be included in the current collection of production and sales data and emissions calculations [11] (see also the chapter on the Production, Sales and Emissions of Indutrial Halocarbeons in this volume). The fluorocarbons in the second group, with the exception of trifluoromethane, HFC-23, all fail to meet the criteria for data collection on the grounds of the limited number of producers, the scale of production, or both. HFC-23 and carbon tetrachloride actually fall outside the scope of the collection of production data; the first is a by-product of HCFC-22 manufacture and the extent to which it is emitted into the atmosphere depends mainly on local environmental legislation, the second is a raw material from which CFCs 11 and 12 are manufactured and historic emissions appeared to have been principally fugitive losses from such processes.

The materials in the third group have either been found to have inadequate properties or are still in the early stages of development.

2
Technical Requirements and Environmental Properties

In addition to having volatility appropriate to the application, there are a number of thermochemical properties that should be optimised. For example, a good thermal insulant gas for blowing plastic insulating foam should have low thermal conductivity, generally associated with relatively high molecular weight; refrigerant and solvent gases should be miscible with oils but not such good solvents that they dissolve parts of the equipment; and for refrigerants operating on the vapour compression cycle (which covers all but a few applications), the triple point should be far above operating conditions. Common to all applications, however, the material should be non-toxic, non-flammable and have an acceptably short environmental lifetime so that it does not cause problems by accumulating in a particular environmental compartment. The latter requirements and optimal thermochemical properties are not necessarily mutually exclusive but they do interact and the choice of material is necessarily a compromise.

Closer inspection of the range of compounds that can be made using the elements identified by Midgley in 1937 [2], in the light of what is now known about environmental effects, tends to reduce the pool of source elements further. Bromine is ruled out on environmental grounds because of the high ozone depletion potential that it can confer; chlorine is less problematic in this respect but can give rise to high ozone depletion if the compound has sufficient atmospheric lifetime. Sulphur compounds are generally likely to be too toxic, leaving C, H, O, N, F and, possibly, Cl as the only viable elemental building blocks now. Ammonia (NH_3) is a well used refrigerant, where safety can be maintained; water vapour was first used at the beginning of this century [1] and, representing species containing C, H, O and F, the hydrofluoroethers have attracted some attention as refrigerants but are still at an early stage of evaluation [12, 13]. However, in the 1980s when alternatives to CFCs were first sought, the group of compounds that appeared to offer the best compromise between safety (both for consumers and for the environment at large) and performance, including initial cost of equipment, comprised the compounds formed of C, H, F and Cl: hydrofluorocarbons (HFCs) and hydrochlorofluorocarbons (HCFCs). Even in this restricted pool, there is a balance among the undesirable properties of toxicity, flammability and environmental lifetime. This was discussed by McLinden and Didion [14], who illustrated the balances using a ternary diagram as shown in Fig. 1.

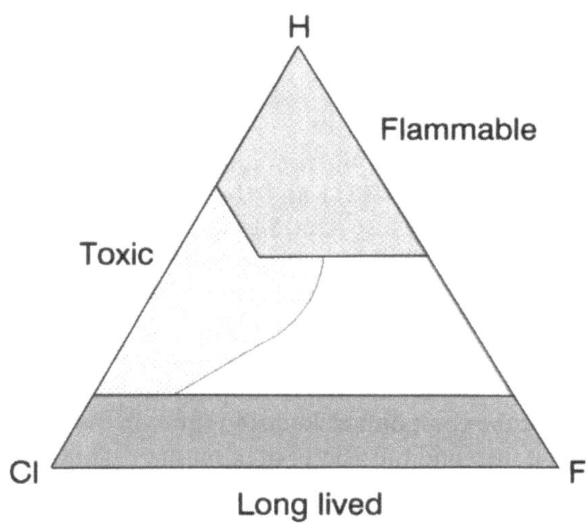

Fig. 1. Balances among flammability, toxicity and atmospheric lifetime for methane to butane derivatives and the dependence on molecular content (as hydrogen, chlorine and fluorine) after McLinden and Didion [14]

The figure illustrates the properties of saturated molecules with one to four carbon atoms, with the rest made up from the elements labelled at the vertices; thus the apex of the triangle would represent CH_4, C_2H_6, C_3H_8 or C_4H_{10}, all hydrocarbons and all highly flammable. Flammability decreases as the halogen content increases, so that the hydrofluorocarbon 1,1,1-trifluoroethane (CH_3CF_3, HFC-143a), which is just flammable, defines the boundary condition on the hydrogen/fluorine axis. The next member, 1,1,1,2-tetrafluoroethane (CF_3CH_2F, HFC-134a) is not flammable under normal conditions.

While the hydrogen content dominates flammability, chlorine content influences both volatility (the more chlorine, the less volatile the material) and toxicity. Members of the group that contain only C, H and F are not toxic. Even with the addition of chlorine to the molecule, most of the hydrochlorofluorocarbons (HCFCs) listed in Table 1 have also been shown to have very low toxicity, with recommended workplace exposure limits that are appropriately high [15]. The exception is 1,1,1-trifluoro-2,2-dichloroethane (HCFC-123) for which manufacturers have set exposure limits of 10–30 ppm and which serves to define the boundary in the middle of the ternary diagram.

Along the base of the triangle are the chlorofluorocarbons (CFCs), containing only C, Cl and F. With no hydrogen in the molecule, these are not susceptible to atmospheric oxidation and hence have long atmospheric lifetimes. Within this sub-group, the lifetimes increase with increasing fluorine content; carbon tetrachloride has an atmospheric lifetime of 42 years [16], but that of carbon tetrafluoride is 50,000 years and perfluoropropane (PFC-218), which is used in some refrigerant blends in small amounts, has a lifetime of 2600 years [6].

The balances among environmental, societal, technical and economic considerations govern the possible choices of replacement refrigerants, foam blowing agents and solvents but they do not inform about the extent of this replacement now, nor about the way that demand could change in the future.

3 Underlying Trends in Demand

For some of the compounds listed in the first group of Table 1, notably CFC-11, CFC-12, methyl chloroform, Halon-1211 and Halon-1301, production data are available right from the start (see also Sect. 5 in the chapter on the Production, Sales and Emissions of Industrial Halocarbons in this volume). Figure 2 shows, on a common basis, the way that global production of these substances built up in time. For each compound there was a year in which production reached a maximum, subsequently declining due to environmentally driven controls (generally the Montreal Protocol); the maximum, and the year in which it occurred, is point 1,1 for all of the compounds. Reduced time, on the x-axis, is the fractional year between the start and the year of maximum production; in true units the scale would have run from 1931 to 1974 for CFC-12 or 1963 to 1988 for the halons. Reduced production, on the y-axis, is similarly the annual production expressed as a fraction of the maximum. Methyl chloroform has the largest actual maximum production and, on the same graph, the line for a constant compound growth rate that fits the minimum and maximum points for methyl

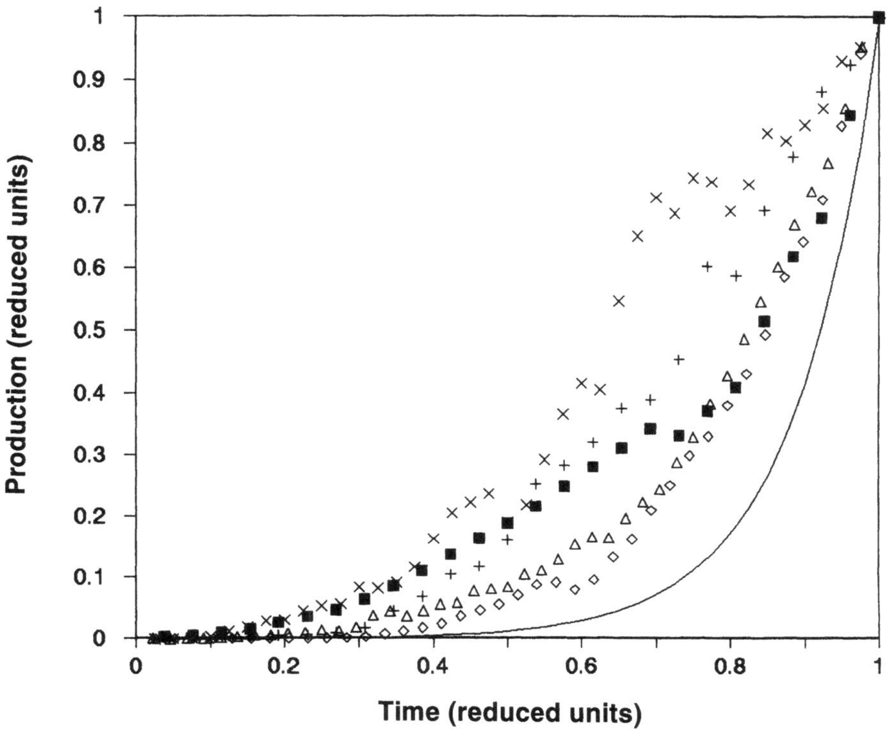

Fig. 2. Growth in production of five halocarbons from beginning to maximum. The process to calculate the reduced units is described in the text. *Open diamonds* CFC-12; *open triangles* CFC-11; *filled squares* Halon-1211; *plus signs* Halon-1301; *crosses* methyl chloroform

chloroform is also shown. It is clear from Fig. 2 that the widely used assumption of compound growth does not fit the observed trends for any of the compounds; a more rigorous analysis is required.

The long-term data sets for CFC 12 and HCFC 22 demand in refrigeration, which were assembled and maintained by industry [11, 17, 18] have been analysed to ascertain the underlying market trends [19, 20].

The thesis used was that the market that is satisfied by materials like CFCs is finite and would eventually reach some upper limit. The magnitude of the limiting value and the rate at which it is reached are governed inter alia by technical parameters such as obsolescence and economics – the costs of using the materials compared to achieving the desired result using another material or technology. There are clear parallels with biological systems that have finite potential for growth, and the underlying trend is an S-curve [21]. The treatment is robust even if there are a number of discrete colonies (or markets) growing at different rates, because the mathematical description of the combined growth is identical to the sum of the individual components. Thus for refrigeration, for example, the growth in both developed and developing countries is included in the treatment so that the curve would continue to describe the market after demand in the developed world has become saturated.

In the case of CFC 12, there are data for almost 40 years of use in refrigeration and air conditioning prior to the introduction of the Montreal Protocol and hence unconstrained by specific regulations. This has been fitted to an S-curve of the form:

$$\log_{10}P = \log_{10}A + k^i \times \log_{10}B \qquad (1)$$

where P is the demand for CFC 12 in all forms of refrigeration and air conditioning in year i (year zero is 1946 to avoid market distortions arising from World War II). If P is in kt (thousands of metric tonnes), A is 616, k is 0.965 and B is 0.0115. R^2 for this line is 0.989. This form of curve is well known within the chemical industry (see for example [22]), and has been shown to describe the historic demand for methyl chloroform [23]. An extrapolation of the line will give an estimate of the underlying future demand for a refrigerant with properties like CFC-12. The result, together with the estimated uncertainty at 95%,

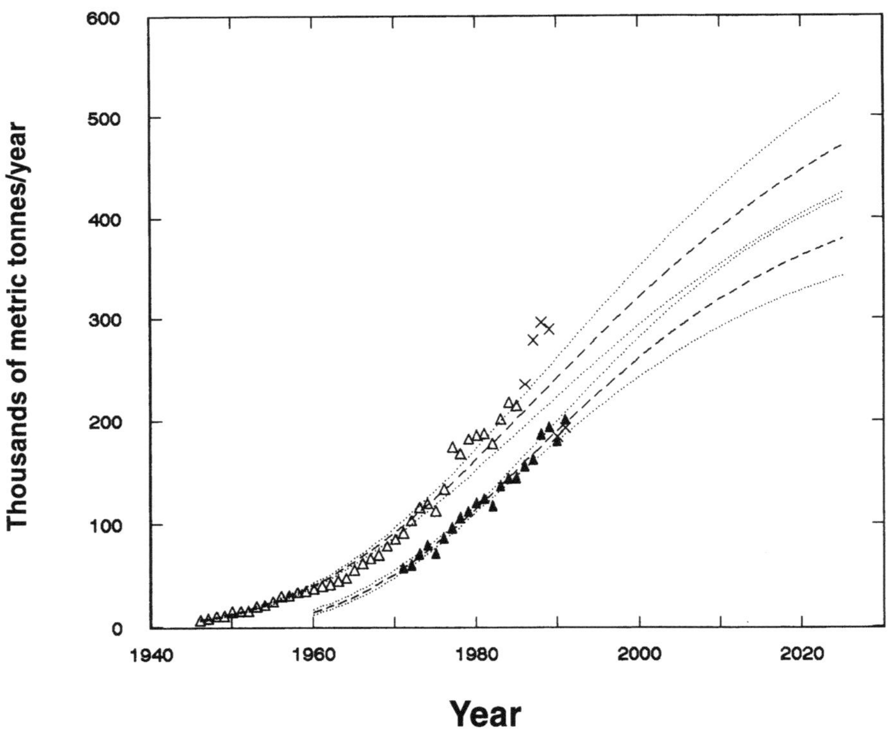

Fig. 3. Growth of CFC-12 and HCFC-22 in refrigeration globally. CFC-12 is represented by *open triangles*, HCFC-22 by *filled triangles*. The fitted lines described in Eq. (1) are represented by *dashed lines* and their 95% confidence limits by *dotted lines*. Points not used in the fitting (see text) are shown as *crosses*

is shown in Fig. 3. It should be noted that the annual consumptions for 1986 to 1991 were not included when calculating the underlying trend; in those years the market was no longer unconstrained and stockbuilding in the advance of regulations, together with the effect of the regulations themselves, resulted in significant distortion.

A similar analysis of the historic HCFC 22 demand reported by AFEAS conveys the same sort of picture [11]. In this case, the values in Eq. (1) are 450 for A, 0.955 for k and 0.0327 for B, with R^2 equal to 0.982, and with i equal to zero in the year 1960. Figure 3 shows the historic data points and projected demand to 2025, together with the 95% confidence limits.

Analyses of this form enable the development of robust scenarios for the total future demand for an effect (such as refrigeration) but do not assess the extent to which this demand would be met by substitute materials or technologies.

4 Substitution

The scope for substitution, even between CFCs, is very limited. Alternative compounds must match the properties needed to provide the specific effect desired and, additionally, should be toxicologically and environmentally acceptable in use. Alternative technologies, on the other hand, have no need to match the physical and chemical properties of CFCs (although they too should be toxicologically and environmentally acceptable) and it is expected that almost half of the demand will be taken up by chemicals other than fluorocarbons or by other technologies [24].

The current indications are that substitution by fluorocarbons actually will be very much less than 50%. In an analysis of the consumption of these compounds in the European Union in recent years, McCulloch and Midgley showed that substitution of CFCs by HCFCs was 26% and the HFC share was only 6% [25]. Figure 4 shows the way that CFC, HCFC and HFC use has changed in the European Union during the period 1986–1996; the total requirement for CFCs of over 300,000 tonnes/year has reduced to zero, with the demand not satisfied by them being met contemporarily by other agents, or other technologies or by better recovery and recycling. The HCFC share of this substitution, at 80,000 tonnes/year, and the HFC share, at 18,000 tonnes/year, are far removed from the near total replacement that has been advocated in some scenario calculations [26] and, indeed, the real rate of substitution by HCFCs in Europe is close to the 25% projected from market research in the early 1990s, while the real rate of substitution by HFCs is very much smaller than the 15% expected in the same work [27]. Global values derived from AFEAS [11] are a better match with the total predicted by McFarland and Kaye [27] but substitution by HFCs is actually lower than predicted (at 11%) and substitution by HCFCs is higher (33%).

In general, the models to describe substitution concentrate on the technical feasibility of replacements, with less importance being attached to the non-technical considerations: price and toxicology. Externalities, such as customer confidence in and perception of the final product, cannot be considered in any but the most subjective manner [28]. On the other hand, announcements of plant

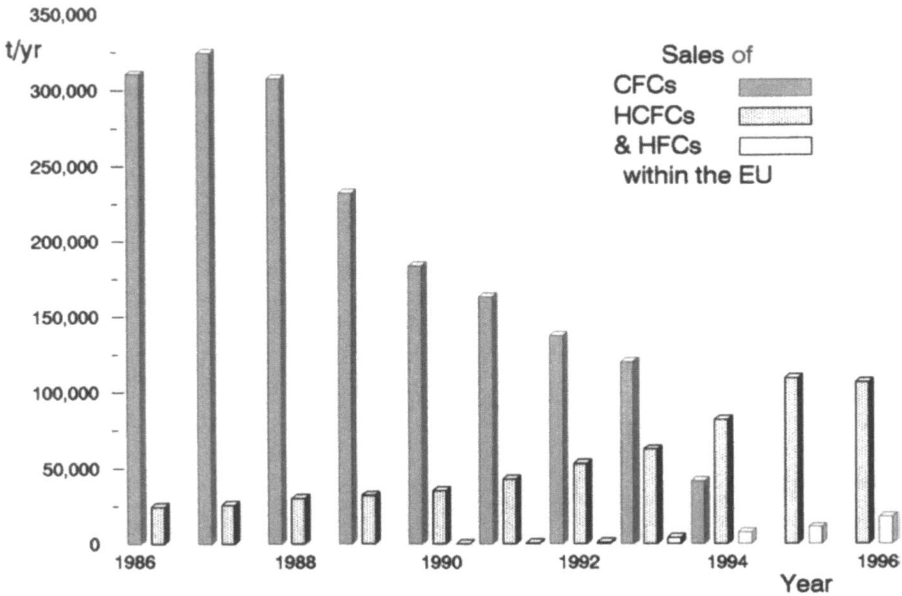

Fig. 4. Annual sales into the European Union of CFCs, HCFCs and HFCs

construction programmes, such as those in the trade press [29], give a clear indication of which alternatives are perceived to be viable. It would seem that the alternatives currently favoured globally are HCFCs 22, 141b and 142b and HFCs 134a and 32, with minor amounts of HCFC-123 and HFCs 125 and 143a [29–31]. The minor compounds have properties which will militate against widespread deployment. HFCs 125 and 143a have atmospheric lifetimes which are comparable with those of the CFCs, giving rise to concerns about persistence in the environment. HFC-152a is flammable and HCFC-123 has adverse toxicological properties. As for HCFC-124, its physical properties do not readily suit it to refrigeration applications and none of the references cited above predicts a significant demand for this compound. This is reflected in the reported global production of HCFC-124, up to the year 1996, being less than 10,000 tonnes in total [11].

The substitutes for the principal uses of CFCs and halons and the demands for halocarbon alternatives in the next few years are summarised in Table 2. While there are many technologies and, indeed, fluorocarbons which have been claimed to provide the effects shown, only those which are technically viable now, and hence may be expected to be commercially available in the near future, are considered. In the longer term it is expected that the needs identified in Table 2 will be met by a variety of techniques and technologies. Fluorocarbon-based technologies are expected to play an important, but not necessarily predominant, rôle.

Considerable quantities of CFCs were used as propellants in aerosol sprays. The vapour pressure of CFC-12 is higher than atmospheric at room temperature,

contained, will operate using HFCs, ammonia, carbon dioxide or hydrocarbon refrigerants.

CFC-113 was used extensively to clean electronic components during the assembly process. The parts of this market that required high precision have adopted HCFC-141b, although there are some water-based systems, and considerable success has been attributed to engineering out the need to clean components at all [9]. Contrary to expectations, the engineering industry has not switched to chlorinated hydrocarbon solvents; demand for these is static or falling and the general trend is towards water-based systems [34].

The principal use for CFC-11, especially in Europe, was to blow plastic insulating foam. This is, essentially, a matrix of polyurethane closed cells trapping the blowing agent, which also acts as the insulating gas. Recent developments of the process first of all enabled the same insulating efficiency with less gas and then led to use of gases such as HCFC-141 b in polyurethane and HCFC-142 b in expanded polystyrene [33]. Hydrocarbons, such as cyclopentane, and water vapour or carbon dioxide generated in situ can also be used to expand the foam and the poorer intrinsic insulating properties, brought about by the higher gas thermal conductivity, compensated by increased thickness. In the longer term, the diminished market where there remains a real need for efficient blown-foam insulation is likely to be met partly by HFC-134a and partly by other HFCs; judging on volatility alone, HFC-245ca would seem suitable. For the rest of the insulation market a variety of technologies (such as vacuum panels and CO_2 or water vapour blowing) are emerging [32, 33].

Halons are different; although halon applications are all intrinsically emissive, with the fire extinguishing action of the agent depending on its emission into the environment, the "banks" in existing equipment, such as fire extinguishers and fixed (total-flood) installations, are much larger multiples of the annual demand than is the case for CFCs [35]. The consequence is that, at present, the principal source of halons is existing equipment at the end of its service life. It is estimated that the halon bank is being depleted at the rate of 11% per year in the case of Halon-1211, and 4% per year for Halon-1301 [36]. Additions to the bank from production in Article 5 countries have had only a small effect and are likely to remain small in the future, given the schedule negotiated for rebate under the Multilateral Fund [37]. Thus, if most of the halon in equipment can be recovered and reused, there is sufficient to last well into the next century. In the meantime, alternative technologies are being improved to offer effects as clean as those of the halons. Alternative halocarbons start with the disadvantage that they cannot contain bromine, which is by far the most effective flame suppressor of the three lighter halogens (F, Cl and Br). It is therefore expected that neither HFC-227ea nor HFC-23 will reach the same level of deployment as Halons 1211 and 1301. Iodine-containing compounds, such as CF_3I, have been suggested as halon replacements; the problem with these is the intrinsic toxicity of iodine compounds.

It is clear that there is no single replacement for each CFC and that other materials and technologies play a variety of roles in substitution, which acts against wholesale adoption of fluorocarbons (HCFCs, HFCs and PFCs) as tonne-for-tonne replacements. The quantities of CFCs that had been used in each application are described in Production, Sales and Emissions of Industrial Halocarbons (this volume) and, when forecasts were made in the early 1990s for demand for HCFCs and HFCs in the latter part of this decade, the values derived were considerably less than those for the CFCs replaced. It was shown earlier in this chapter that the projections for HCFCs by McFarland and Kaye in 1992 were remarkably prescient [27], while those for HFCs, at least in Europe, were rather too high. This is reflected in the projections for individual compounds, such as those made by McCarthy in 1993 who suggested 200 kt/year global demand for HCFC-22 (actual 230 kt/year) and 140 kt/year as the demand for HCFCs used in solvent and foam blowing applications (actual 152 kt/year) [31]. Global demand for HFC-134a in refrigeration and air-conditioning was expected to be in the region of 180 kt/year [31, 38] but, so far, only 71 kt/year of that demand has materialised [11].

5 Future Scenarios

5.1 Ozone Depleting Substances

The accumulation of ozone depleting substances in the atmosphere, and the resulting chlorine and bromine loading of the troposphere, is well recorded by atmospheric measurements [16, 39] and it is possible to balance recorded emissions with these atmospheric concentrations after taking account of environmental loss processes [40]. Over the past few years, the rise in chlorine and bromine loading has slowed, generally as expected from the controls imposed by the Protocol [39, 41] and it is apparent that we have now passed a peak in chlorine loading, but that bromine loading continues to increase.

Using the current control measures of the Montreal Protocol shown in Table 1 of the chapter on International Regulation of Halocarbons (this volume), it is possible to construct scenarios that show the maximum allowed production of CFCs, HCFCs, carbon tetrachloride, methyl chloroform, halons and methyl bromide during the next century. As described in the latest assessment [39], these scenarios are important to enable investigation of the sensitivity of future chlorine and bromine loadings to changes in the control measures. In previous assessments [16, 42], it had been sufficient to assume production, consumption and emissions were equal and contemporaneous which was adequate to show gross changes over a long timescale. As production decreases, however, it is necessary to take account of the fact that the materials are used before being emitted. Only relatively small amounts of the materials escape during manufacture [17]; the principal emissions occur during and after use of the fluids in, for example, aerosols, refrigeration, foam blowing and solvent applications. Consequently, emissions are estimated as the product of an emission function and the quantity of material that remains in service in equipment. This quantity is normally referred to as the "bank" of unemitted material and, arithmetically, is simply the difference between the total quantity of the substance manufactured and the quantity that has been emitted. Thus an accurate estimate of the bank underlies calculations of quantity and longevity of future emissions.

As described in the chapter on the Production, Sales and Emissions of Industrial Halocarbons (this volume), historical production and releases of materials are documented in the industrial surveys now published by the Alternative Fluorocarbons Environmental Acceptability Study [11]. These cover production in facilities in OECD countries (and those factories owned by the same manufacturers in the rest of the world) of CFCs 11 and 12 from 1931 to the present day, and CFCs 113, 114 and 115 from 1970 to now. Production in the rest of the world during recent years, described in the industrial database as "non-reported" production, can be estimated using the values given for individual nations in the UN statistics for the Montreal Protocol [43]. As discussed by Oberthür [44], the coverage and quality of this database are insufficient to provide an historical record of the same standard as the industrial data; however, comparison of the two data sets led to the conclusion that the non-reported production could be represented by a constant 120,000 ODP tonnes over the period 1989–1995. In the study recently carried out for the 1998 Assessment of Ozone Depletion [39], this quantity was subdivided among the CFCs and added to the production reported by AFEAS [11].

Future production in the CFC scenario comprises the allowance for production in Article 5 countries (Sect. 2.2 in the chapter on International Regulations on Halocarbons in this volume), plus 15% of the quantity that industrialised countries produced during 1986 (another allowance for consumption by Article 5 countries), plus additional production for "essential" uses, as defined by the Parties to the Protocol in, for example, [45]. This scenario gives the maximum credible concentrations from emissions that result from production at the highest levels allowed under the Montreal Protocol. It does not include emissions from material traded illegally, nor does it include specific reductions in CFC production claimed for rebates under the Multilateral Fund (see Sect. 2.2 in the chapter on International Regulations on Halocarbons in this volume).

The production and consumption of other fully halogenated CFCs, those covered by Annex B-I of the Protocol, was less than 3400 tonnes in 1989, the reference year, and had fallen to 64 tonnes by 1995 [43]. The 1989 production was less than 0.4% of the value for the Annex A-I substances listed above and so the other CFCs were not considered further.

Because the scenario does not discriminate end-uses, it was not possible to calculate emissions in the way described in Sect. 3 of the chapter on the Production, Sales and Emissions of Industrial Halocarbons (in this volume), and so emission functions for the future were assigned as simple, constant fractions of the remaining banks of material in equipment and systems, at values based on historical data. Past emissions of CFCs were calculated by inverse modelling from the reported atmospheric concentrations [46, 47] using a single box model and the atmospheric lifetimes in Table 1. The average fraction of the remaining bank represented by these emissions was then applied for all years up to 2100 [11].

Similar treatments yielded estimates for future emissions of methyl chloroform (where historic production and emissions were taken from [23, 48, 49]) and halons (where the sources of historic data were [35, 36]). During the 1980s, emission of carbon tetrachloride was well represented by approximately 8% of the production of CFC-11 and CFC-12 [50] and so this function was used for the future, implying that, after complete phase out of CFCs in 2010, carbon tetrachloride emissions will cease.

Methyl bromide emissions were based on the treatment expounded in the latest report from SORG [41] that showed that a significant fraction of the methyl bromide entering the atmosphere was not accountable to known natural and anthropogenic sources. The small fraction that the Montreal Protocol controls was reduced accordingly and the remainder kept constant.

Up to 1990, only one HCFC, chlorodifluoromethane (HCFC-22), was emitted in significant quantities; subsequently, HCFCs 141b and 142b have been released into the environment in sufficient quantities to be measured in the atmosphere [46, 51]. Production and sales data are collected for these compounds. HCFC-124 was added to the list this year after the scenarios had been developed for WMO [11]; consequently, the scenarios do not contain HCFC-124 although they do contain HCFC-123. In either case, the environmental effects are small; quantities of HCFC-124 are small and the atmospheric lifetime of HCFC-123 is too short to allow accumulation on any sensible emission scenario. The future consumption of HCFCs is less predictable than that of the CFCs, due to the fact that controls, in the form of the cap on consumption, apply only to developed countries. Article 5 countries could, if they desired, allow their production to grow continuously until 2015. The total scenario adopted for HCFCs was the revised IS92a reference scenario in the Second Assessment Report of IPCC [6], with the difference between the cap and the quantity envisaged in IS92a being assigned to the Article 5 countries. By 2015, this notional production in the developing world would have grown at an average rate of 4.3% per year to reach a value that is 17% of the maximum that had been produced in Article 2 countries. Even though the choice of HCFC scenario is somewhat arbitrary, the growth is actually greater than the growth in GDP projected over the same period and the result seems reasonable [26].

The complete scenario, which was used to calculate the Equivalent Effective Stratospheric Chlorine concentrations in Fig. 1 of the chapter on International Regulation of Halocarbons (this volume), is summarised in Table 3.

5.2 Halocarbon Greenhouse Gases

Leaving aside the ozone depleting substances that are also greenhouse gases, currently the highest concentrations of other halocarbon greenhouse gases in the atmosphere result from by-product emissions, rather than from the release of materials manufactured deliberately. Tetrafluoromethane (carbon tetrafluoride, CF_4), with lesser amounts of hexafluoroethane (C_2F_6), is released from primary aluminium production during "anode effect" events in the Hall electrolytic process. In the normal situation, alumina (Al_2O_3) dissolved in cryolite (Na_3AlF_6) is electrolysed to aluminium and oxygen. Under abnormal conditions, however, the cryolite reacts with the carbon cell anode and yields tetrafluoromethane and hexafluoroethane [52]. Currently, the atmosphere carries some 75 ppt of CF_4 and 2.5 ppt of C_2F_6 and, from the rate of growth of these concentrations, Harnisch et al. calculated emissions to be 14,600 tonnes/year of CF_4

Table 3. Emissions of ozone depleting substances in the Montreal Protocol scenario (kt)

Year	CFC-11	CFC-12	CFC-113	CFC-114	CFC-115	CCl_4	CH_3CCl_3	HCFC-22	HCFC-141b	HCFC-142b	HCFC-123	Halon-1211	Halon-1301	Halon-2402	CH_3Br
1990	250	371	236	10.3	12.2	63	718	178	0	9	0	10.1	4.3	1.8	206
1995	105	173	48	3.1	10.9	40	234	183	46	29	0	7.3	2.8	1.8	206
2000	97	181	47	2.2	2.2	24	0	350	65	49	125	6.5	2.7	0.6	201
2005	87	179	44	1.0	0.4	19	0	432	57	54	97	6.3	2.6	0.2	181
2010	72	133	16	0.5	0.1	0	0	412	49	47	58	4.9	2.3	0.1	181
2015	45	31	0	0.2	0.0	0	0	283	33	31	26	2.7	1.9	0.0	167
2020	28	7	0	0.1	0.0	0	0	137	16	15	16	1.5	1.5	0.0	167
2025	15	2	0	0.0	0.0	0	0	77	10	9	16	0.9	1.3	0.0	167
2030	11	0	0	0.0	0.0	0	0	66	9	8	16	0.5	1.0	0.0	167
2035	7	0	0	0.0	0.0	0	0	63	8	8	16	0.3	0.8	0.0	167
2040	4	0	0	0.0	0.0	0	0	62	8	8	16	0.1	0.7	0.0	167
2060	1	0	0	0.0	0.0	0	0	0	0	0	0	0.0	0.3	0.0	167
2080	0	0	0	0.0	0.0	0	0	0	0	0	0	0.0	0.1	0.0	167
2100	0	0	0	0.0	0.0	0	0	0	0	0	0	0.0	0.1	0.0	167

and 1900 tonnes/year of C_2F_6 [53]. On a global aluminium production of 20 million tonnes/year in the early 1990s, Weston calculated the emission of CF_4 to have been between 26,000 and 32,000 tonnes/year by using a higher emission factor [54]. Both sources agree that a large proportion (of the order of 60%) of the tetrafluoromethane present in the atmosphere is not the result of man's activities. Nevertheless, the growth in concentration is almost wholly due to aluminium production which is expected to reach 30–40 million tonnes/year by the end of the next century. With the higher emission factor from Weston [54], this could lead to an atmospheric concentration of CF_4 in the region of 600–900 ppt, if the emission per tonne of aluminium remained constant. The effect, however, can be drastically reduced by better design of the cell and its control system and so anode effect events are more common with older cells; it is likely that the actual concentration in the future will be less than that estimated. In the calculation of emissions and atmospheric concentration reported below, the lower emission factor of Harnisch et al. is used [53].

HFC-23 (trifluoromethane, CF_3H) is a by-product arising from overfluorination during the manufacture of HCFC-22 (chlorodifluoromethane). In the past some was recovered to be used as feedstock for Halon-1301 manufacture, but this has now all but ceased and, because it is both non-toxic and non-flammable, a common means of disposal is still to vent the material to atmosphere. Oram et al. report growth in the atmospheric concentration from 3 ppt in 1978 to 10 ppt in 1996 [55] and, coupled with a global total production of HCFC-22 calculated from AFEAS [11] and Midgley and Fisher [18], estimated an emission factor of 21 kg of HFC-23 per tonne of HCFC-22. It is to be expected that this factor will decrease in the future as more and more regulatory authorities require emission of HFC-23 to be minimised purely on the grounds of pollution control. The only significant use for HFC-23 is in total flood fire extinguishing installations where it is in competition with carbon dioxide and inert gas and water systems. Demand is never likely to be as high as for Halon-1301. Furthermore, this is a dispersive use – the material deployed is eventually emitted to atmosphere – so that recovery of HFC-23 from HCFC-22 plants for use in fire fighting has the same effect on climate in the long term as if it had been released from the plants. For the future, HFC-23 emissions can be expected to follow total HCFC-22 production; that is the quantity required for the scenario shown in Table 3, plus the quantity required as a feedstock for the manufacture of fluoropolymers (currently of the order of an extra 35% [18]).

HFCs are also released as by-products from other HFC manufacturing processes but the average by-product emissions from HFC-134a production, for example, are insignificant compared with the production of HFC-134a itself and no single compound predominates [56]. These are not considered further.

For the fluorocarbons that are being produced to replace CFCs, the Intergovernmental Panel on Climate Change has produced reference scenarios [5, 6, 26]. These are based on extrapolations of Gross Domestic Product (GDP) provided by organisations such as the World Bank. Fluorochemical usage is estimated using a constant relationship with GDP, the relationship being based on historical GDP and CFC usage, and views on the extent of substitution. This approach gives a growth rate that is very similar to that derived from a market based calculation in the short term, up to say 2020, but extrapolation beyond then is dubious [20]. In addition, all scenarios are highly dependent on sound estimates for substitution; the better estimates are rooted in historical measurements rather than views on suitability for particular functions, which tend to owe more to theory than practice.

As described in Sect. 3 there are clear underlying trends in demand for refrigeration which is likely to be an enduring end-use for these materials. Long term extrapolation of Eq. (1) for both CFC-12 and HCFC-22 shows that the peak requirement for both compounds, together, would have reached 1.1 million tonnes/year by the middle of the next century. Consideration of the historic AFEAS data shows that the total recorded production of all CFCs, for all uses other than as solvent, was 15.4 million tonnes, of which refrigeration used 7.1 million tonnes (or 46%) [11]. Comparison of 1996, the latest year for which there are data, with 1987, the peak year of production, shows a reduction in CFC production and use of 763,000 tonnes/year, again excluding solvent uses that were shown in Sect. 4 to have, in most cases, gone to technologies other than fluorocarbon already. The increment in all HCFC and HFC production and use, over the same time period, was 340,000 tonnes/year (or 44%). Since substitution of CFC uses by alternative agents must be contemporaneous, this value represents the total global extent to which uses formerly served by CFCs are currently being substituted by both HCFCs and HFCs. Other technologies must be substituted for the remainder and it is to be expected that, as a general rule, once another technology has successfully supplanted fluorocarbons in an application, users will not revert to using fluorocarbons in the future. In the EU, the extent of substitution is somewhat lower, amounting to 26% by HCFCs and 6% by HFCs but the global values are used in this estimate [25]. Since the fraction of fluorocarbon substitution and the fraction of the demand represented by refrigeration are very similar, the demand pattern for CFC-12 and HCFC-22, combined, can be used as a guide to the future demand for all fluorocarbons substituting mainly in refrigeration and foam blowing but with minor amounts for aerosols, metered dose inhalers and specialist solvents. Thus the maximum requirement for all fluorocarbons is likely to be in the region of 1 million tonnes/year by the middle of the next century, remaining substantially constant thereafter.

Subdivision of this total into individual end uses and the compounds that could meet the requirements of those uses is a great deal more subjective. As a relatively simple first step, a constant allowance can be made for metered dose inhalers and for the fire-extinguishing applications that must have clean agents of the form offered by fluorocarbons. In recent history, production of Halon-1211 was in the region of 10,000 tonnes/year and an allowance of 10,000 tonnes/year is suggested for metered dose inhalers, based on the essential uses sanctioned for CFCs [45]. Accordingly, a constant 20,000 tonnes/year of demand was allocated to HFC-227ea (or something with similar properties). The quantities of other materials (such as HFC-134a) that are used in these applications will be small compared to their principal use and, for the purposes of a simple estimate, can be accommodated in that.

From the arguments put forward in Sect. 4, it is likely that refrigeration will be a larger fraction of total use than historically. Some 56% of the demand is expected to be in refrigeration and to be served by HFCs 134a, 32, 125 and 143a and PFC-218, or materials with similar technical properties. The remaining 44% is expected to be in foam blowing and to be served by HFCs 134a and 245ca (or, again, compounds with similar properties). HFC-134a is the only one of these compounds that is known to be used in these duties and the only one reported to have been measured in the atmosphere [57], although HFCs 125 and 143a have been observed at about their detection limit (ca. 1 ppt, 1 in 10^{12}, P.G. Simmonds, personal communication). There is no evidence that HFC-32 is accumulating in the atmosphere, so its deployment may be much smaller than envisaged, for example in [20] and, for the purposes of this calculation, HFC-32 is included in the total expected for HFC-134a. These observations would be matched if 5% of the demand were assigned to HFC-125 and another 5% to HFC-143a. Perfluorocarbons (PFCs) other than CF_4 have not been observed in the atmosphere and, due to their cost, are not likely to be used to the same extent as HFCs, so that 2% of the demand has been allocated to PFC. There are no technical data on which to base a choice of compound here and so PFC-218 (C_3F_8) is used as the model because it is the first of the PFC series that is likely to be useful and, having the lowest molecular weight of the rest of the series, will produce the highest atmospheric concentration per unit mass emitted. Some 44% of the total demand is therefore expected to be accountable to HFC-134a (plus HFC-32) used in refrigeration.

For foam blowing, equal amounts of HFCs 134a and 245ca are assumed. Because HCFCs 141b and 142b are still available, comparatively little foam is blown with HFCs; the extent of future demand here is more robustly based than the choice of compounds and it may be that a variety of fluorocarbons will actually be used. However, HFC-245ca would appear to have suitable properties and so should serve as a useful model compound in the scenarios.

The minor uses, in specialist solvents and in aerosol propulsion, are small components of the total and were included in the calculation of its overall size. Also no other HFCs are considered. If there were to be any, they would share part of the market with the model compounds and would not represent an additional emission.

For several years during the interval between now and the development of the maximum demand, both CFCs and HCFCs will satisfy some of that demand, so reducing the requirement for HFCs then. The maximum quantities permitted by the Montreal Protocol is reflected in the emissions in Table 3 but this is as unlikely to be realised as is the scenario in which there are no future emissions. It is probable that much less CFCs will be produced than is allowed and so no future production of CFCs is offset against the demand scenario. In addition, the allowance for HCFC-123 is not included, on the basis that this is closest to the solvent applications that have been lost from the future demand. Both of these tend to augment the quantities of HFCs calculated for the early part of the next century. Having discounted all of the demand otherwise satisfied, the expected emissions of HFCs and PFCs are shown in Table 4. Because the scenarios are not discriminated wholly into individual end uses, emissions functions were calcu-

Table 4. Emission scenario for halocarbon greenhouse gases

Year	HFC-134a (including HFC-32)	HFC-245ca[a]	HFC-125[a]	HFC-143a[a]	PFC-218[a]	HFC-227ea[a]	HFC-23	CF_4
	kt Refrigeration and foam blowing	kt Foam blowing	kt Refriger-ation	kt Refriger-ation	kt Refriger-ation	kt MDIs[b] and fire extin-guishers	kt HCFC-22 by-product	kt Aluminium by-product
2000	94	0	5	5	1	20	10	15
2010	208	69	16	16	6	20	7	18
2020	429	145	33	33	13	20	2	20
2030	570	193	44	44	18	20	1	23
2040	625	212	48	48	19	20	1	27
2050	654	222	50	50	20	20	1	31
2060	657	222	51	51	20	20	1	31
2070	663	224	51	51	20	20	1	31
2080	671	227	52	52	21	20	1	31
2090	678	230	52	52	21	20	1	31
2100	686	232	53	53	21	20	1	31

[a] Indicates that the material is a model compound; the actual material used may be different.
[b] Metered dose inhalers.

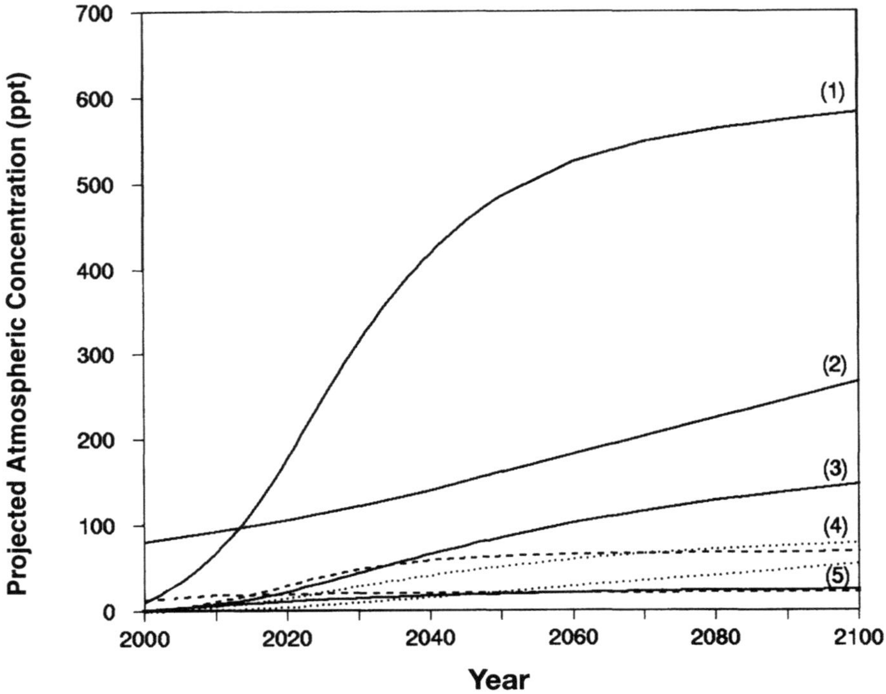

Fig. 5. Predicted future concentrations of HFCs and PFCs in the atmosphere, based on the scenarios in Table 4. *1* HFC-134a (including a small fraction that is HFC-32) from refrigeration and foam blowing; *2* tetrafluoromethane from aluminium production; *3* HFC-143a* from refrigeration; *4 dotted line* HFC-125* from refrigeration, *dashed line* HFC-245ca* from foam blowing; *5 dotted line* PFC-218* from refrigeration, *dashed line* HFC-23 from production of HCFC-22, *solid line* HFC-227ea* from metered dose inhalers and fire extinguishers. * signifies model compounds; the actual material used could be different

lated globally from the data from AFEAS [11]. This shows that an average of 25% of the unreleased "bank" of each of the new fluorocarbons was emitted annually during the period 1992–1996. Accordingly, 25% of the bank was used as a general emission function for all compounds for the whole of the next century. Expected emissions of by-products tri- and tetrafluoro-methane are also shown in Table 4, calculated as described above.

The consequences of these emissions in terms of atmospheric concentrations is shown in Fig. 5. The principal differences between these curves and those derived from the IPCC IS92 scenarios (as described by Harnisch) are in the shape and the absolute magnitudes predicted. As discussed previously in Sect. 3, scenarios that rely on parametrised alignment with GDP will predict a continuing acceleration in growth that is not generally borne out in practice, and the tendency towards stabilisation in both emissions and atmospheric concentrations shown in Table 4 and Fig. 5 have a more realistic basis in history. The higher concentrations arising from the IS92 scenarios are functions of higher emissions that result from total substitution. As shown earlier, there is evidence from both global and regional data that rates of substitution are significantly less than 50% [11,25].

5.3 Other Halocarbons

As described by Aucott et al. [58] and McCulloch et al. [34], anthropogenic emissions of chloroform (trichloromethane, $CHCl_3$), dichloromethane (methylene chloride, CH_2Cl_2), trichloroethene (trichloroethylene, $CHCl=CCl_2$) and tetrachloroethene (perchloroethylene, $CCl_2=CCl_2$) are either static or have fallen in the past few years. Man-made emissions of chloroform arise principally from chlorination of humic material, and the principal sources are paper-making and potable water treatment; current emissions are 70,000 tonnes/year within a total flux of 630,000 tonnes/year [59]. In 1990, dichloromethane, trichloroethene and tetrachloroethene were emitted globally by industry at 580,000, 240,000 and 366,000 tonnes/year, respectively. In the same year, the calculated total fluxes were 840,000, 432,000 and 514,000 tonnes/year, indicating substantial non-industrial sources [59]. For the near future, these anthropogenic inputs into the environment can be treated as constant, with a trend towards reduction in the longer term. Since there are only a few years data and the production of these materials is heavily influenced by economic cycles, it is not possible to characterise the rate of reduction.

6 References

1. Calm JM, Didion DA (1997) Trade-offs in refrigerant selections: past, present and future. In: Refrigerants for the 21st Century. ASHRAE, Atlanta
2. Midgley T Jr (1937) Ind Eng Chem 29:239
3. Hamilton JM Jr (1963) Adv Fluorine Chem 3:117
4. UNEP (1997) The 1987 Montreal Protocol on substances that deplete the ozone layer as adjusted and amended by the 2nd Meeting of the Parties (London, 27–June 1990) and by the 4th Meeting of the Parties (Copenhagen, 23–November 1992) and further adjusted by the 7th Meeting of the Parties (Vienna, 5–December 1995) and further adjusted and amended by the 9th Meeting of the Parties (Montreal, 15 September). UNEP, Nairobi
5. Houghton JT, Meira Filho LG, Bruce J, Lee H, Callander BA, Haites E, Harris N, Maskell K (eds) (1994) Climate change 1994: radiative forcing of climate change and an evaluation of the IPCC IS92 emission scenarios. University Press, Cambridge
6. Houghton JT, Meira Filho LG, Callander BA, Harris N, Kattenberg A, Maskell K (eds) (1996) Climate change 1995: the science of climate change, contribution of working group 1 to the second assessment report of the IPCC. University Press, Cambridge
7. Grasselli JG, Weast RC (eds) (1989) Handbook of data on organic compounds. CRC Press, Boca Raton
8. Hoechst (1997) HFA 227 Pharmaceutical Grade, Safety Data Sheet. Hoechst, Frankfurt
9. ADL (1994) Update on comparison of global warming implications of cleaning technologies using a systems approach. Arthur D. Little, for AFEAS, Washington
10. Decaire BR, Pham HT, Richard RG, Shankland IR (1993) Blowing agents the next generation. In: Proc Global Business Outlook for CFC Alternatives. Intertech, Portland

11. AFEAS (1998) Production, sales and atmospheric release of fluorocarbons through 1996. AFEAS, Washington
12. Heathfield AE, Anastasi C, Pagsberg P, McCulloch A (1998) Atmos Environ 32:711
13. Heathfield AE, Anastasi C, Ballard J, Newnham DA, McCulloch A (1998) J Quant Spectrosc Radiat Transfer 59:91
14. McLinden MO, Didion DA (1987) ASHRAE Journal 29:32
15. PAFT (1997) Testing to extremes. PAFT, Washington
16. WMO (1994) Scientific assessment of ozone depletion: 1994. WMO, Geneva
17. Gamlen PH, Lane BC, Midgley PM, Steed JM (1986) Atmos Environ 20:1077
18. Midgley PM, Fisher DA (1993) Atmos Environ 27A:2215
19. McCulloch A (1994) Environ Monitor Assess 31:167
20. McCulloch A (1995) Environ Int 21:353
21. Norton L, Simon R, Brereton HD, Bogden AE (1976) Nature 264:542
22. Luker BG (1961) Br Plastics 108
23. Midgley PM (1989) Atmos Environ 23:2663
24. Hurley BJ (ed) (1993) Global environmental change report (GECR), vol 1. Cutter Information, Arlington
25. McCulloch A, Midgley PM (1998) Atmos Environ 32:1571
26. Houghton JT, Callander BA, Varney SK (eds) (1992) Climate change 1992: the supplementary report to the IPCC scientific assessment. University Press, Cambridge
27. McFarland M, Kaye J (1992) Photochem Photobiol 55:911
28. Shapiro MM (1993) Substitution assumptions. In: Background papers to workshop on other greenhouse gases – HFC and PFC emissions. USEPA, Washington
29. ECN (1993) Eur Chem News Process Rev p13
30. Verhille M (1992) Kommission-drucksache 12/6f. In: Testimony to the Enquete-Kommission of the Deutscher-Bundestag, "Schutz des Menschen und der Umwelt". Deutscher-Bundestag, Bonn
31. McCarthy R (1993) CFC alternatives: an overview. In: Proc Global Business Outlook for CFC Alternatives. Intertech, Portland
32. Sand JR, Fischer SK, Baxter VD (1997) Energy and global warming impacts of HFC refrigerants and emerging technologies. AFEAS, Washington
33. Fischer SK, Tomlinson JJ, Hughes PJ (1994) Energy and global warming impacts of not-in-kind and next generation CFC and HCFC alternatives. AFEAS, Washington
34. McCulloch A, Aucott ML, Graedel TE, Kleiman G, Midgley PM, Li Y-F (1998) J Geophys Res (in press)
35. McCulloch A (1992) Atmos Environ 26A:1325
36. Fraser PJ, Oram DE, Reeves CE, Penkett SA, McCulloch A (1998) J Geophys Res (submitted)
37. UNEP (1997) China's halon sector strategy, a report by the Executive Committee of the Multilateral Fund. Document No. UNEP/OzL.Pro/ExCom/23/11. UNEP, Nairobi
38. GBEC (1994) Responsibility for the future: options for sustainable management of substance chains and material flows. Fehlberg W, Ulloa-Fehlberg M (trs) Economica, Bonn
39. WMO (1998) Scientific assessment of ozone depletion: 1998. WMO, Geneva
40. Prather M, Midgley P, Rowland FS, Stolarski R (1996) Nature 381:551
41. SORG (1996) Stratospheric ozone 1996. UK Department of the Environment, London
42. WMO (1991) Scientific assessment of ozone depletion: 1991. WMO, Geneva
43. UNEP (1997) Production and consumption of ozone depleting substances, 1986–1995. UNEP, Nairobi
44. Oberthür S (1997) Production and consumption of ozone depleting substances 1986–1995. Deutsche Gesellschaft für Zusammenarbeit, Eschborn
45. UNEP (1997) Quantity of controlled substances authorized under the essential-use process. In: Issues before the open-ended working group in its 15th meeting. Document No. UNEP/OzL.Pro/WG.1/15/2/Add.7. UNEP, Nairobi
46. Montzka SA, Butler JH, Myers RC, Thompson TM, Swanson TH, Clarke SD, Locke LT, Elkins JW (1996) Science 272:1318

47. Prinn RG, Weiss RF, Fraser PJ, Simmonds PG, Alyea FN, Cunnold DM (1998) The ALE/GAGE/AGAGE database, dataset No DB-1001. DOE-CDIAC, Oak Ridge
48. Midgley PM, McCulloch A (1995) Atmos Environ 29:1601
49. Midgley PM, McCulloch A, Simmonds PG (1998) The phase-out of methyl chloroform and the decline in tropospheric chlorine. In: Proc EUROTRAC-2 Symposium, Transport and Chemical Transformation in the Troposphere. Garmische-Partenkirchen
50. Simmonds PG, Cunnold DM, Weiss RF, Prinn RG, Fraser PJ, McCulloch A, Alyea FN, O'Doherty S (1998) J Geophys Res 103:16,017
51. Oram DE, Reeves CE, Penkett SA, Fraser PJ (1995) Geophys Res Lett 22:2741
52. Calandra AJ, Castellano CE, Ferro CM, Cobo O (1982) Experimental and theoretical analysis of the anode effects in industrial cells. In: Light metals. Metallurgical Soc America, Washington, p 345
53. Harnisch J, Borchers R, Fabian P, Maiss M (1996) Geophys Res Lett 23:1099
54. Weston RE Jr (1996) Atmos Environ 30:2901
55. Oram DE, Sturges WT, Penkett SA, McCulloch A, Fraser PJ (1998) Geophys Res Lett 25:35
56. McCulloch A, Campbell NJ (1998) Climate change implications of producing refrigerants. In: Proc Natural Working Fluids Conference. International Institute of Refrigeration, Paris
57. Oram DE, Reeves CE, Sturges WT, Penkett SA, Fraser PJ, Langenfelds RL (1996) Geophys Res Lett 23:1949
58. Aucott ML, Graedel TE, Kleiman G, McCulloch A, Midgley PM, Li Y-F (1998) J Geophys Res (in press)
59. Keene WC, Khalil MAK, Erickson DJ III, McCulloch A, Graedel TE, Lobert JM, Aucott ML, Gong S-L, Harper DB, Kleiman G, Koropalov V, Midgley PM, Moore RM, Seuzaret C, Sturges WT, Barrie LA, Logan JA, Li Y-F (1998) J Geophys Res (in press)

CHAPTER 6

Production, Sales and Emissions of Halocarbons from Industrial Sources

P.M. Midgley[1] and A. McCulloch[2]

[1] M&D Consulting, Ludwigstraße 49, D-70771 Leinfelden, Germany.
E-mail: paulinemidgley@compuserve.com
[2] ICI Chemicals & Polymers, Safety & Environment Department, P.O. Box 13, The Heath, Runcorn, Cheshire WA7 4QF, UK. *E-mail: archie_mcculloch@ici.com*

Calculations of global emissions of atmospheric trace species of industrial origin require accurate information on their production and sales. In this Chapter, data are presented on the production, sales and emission of chlorofluorocarbons, halons, hydrochlorofluorocarbons, hydrofluorocarbons, and the following chlorinated solvents: methyl chloroform, methylene chloride, tetrachlorethene and trichloroethene. Data available from the United Nations Environment Programme and from the audited surveys carried out by the halocarbon industry are described. The method of deriving global emissions from these data for the scientific study of the atmosphere is outlined. The most recent information on production, sales and emissions of the industrial halocarbons listed above is reported in a series of Tables and Figures.

Keywords: Chlorofluorocarbons, Halons, Hydrochlorofluorocarbons, Hydrofluorocarbons, Chlorinated solvents, Montreal Protocol

The Handbook of Environmental Chemistry Vol. 4 Part E
Reactive Halogen Compounds in the Atmosphere
(ed. by P. Fabian and O. N. Singh)

Abbreviations

AFEAS	Alternative Fluorocarbons Environmental Acceptability Study
CEFIC	Conseil Européen des Fédérations de l'Industrie Chimique
CFC(s)	chlorofluorocarbon(s)
CFC-11	F-11, $CFCl_3$
CFC-12	F-12, CF_2Cl_2
CFC-113	CCl_2FCClF_2
CFC-114	$CClF_2CClF_2$
CFC-115	$CClF_2CF_3$
CMA FPP	Chemical Manufacturers' Association, Fluorocarbon Program Panel
GEIA	Global Emissions Inventory Activity
GDP	Gross Domestic Product
HBFC(s)	hydrobromofluorocarbon(s)
HCFC(s)	hydrochlorofluorocarbon(s)
HCFC-22	$CHClF_2$
HCFC-123	$CHCl_2CF_3$
HCFC-124	$CHClFCF_3$
HCFC-141b	CH_3CCl_2F
HCFC-142b	CH_3CClF_2
HFC(s)	hydrofluorocarbon(s)
HFC-134a	CH_2FCF_3
HFC-23	CHF_3, fluoroform
Halon	fully saturated compound containing inter alia C, Br and F, and used for fire-extinguishing
Halon-1211	H-1211, CF_2BrCl
Halon-1301	H-1301, CF_3Br
Halon-2402	H-2402, CF_2BrCF_2Br
kt	thousands of (metric) tonnes; $1 kt = 10^3 t = 10^6 kg = 1 Gg$
ODP	ozone depletion potential
ODS	ozone-depleting substance
OECD	Organisation for Economic Cooperation and Development
RCEI	Reactive Chlorine Emissions Inventory
SORG	Stratospheric Ozone Review Group (UK)
TRI	Toxic Release Inventory
UNEP	United Nations Environment Programme
USEPA	United States Environmental Protection Agency
CH_3Cl	methyl chloride (chloromethane)
CH_2Cl_2	methylene chloride (dichloromethane)
$CHCl_3$	chloroform (trichloromethane)
CCl_4	carbon tetrachloride (tetrachloromethane)
CH_2ClCH_2Cl	ethylene dichloride (1,2-dichloroethane)
CH_3CCl_3	methyl chloroform (1,1,1-trichlorethane)
$CHCl_2CHCl_2$	1,1,2,2-tetrachloroethane
$CHCl_2CCl_3$	pentachloroethane
$CClH{=}CH_2$	vinyl chloride (chloroethene)

$CCl_2{=}CH_2$	vinylidene chloride (1,1-dichloroethene)
$CHClCCl_2$	trichloroethene (trichloroethylene)
CCl_2CCl_2	perchloroethene (perchloroethylene)

1 Introduction

Accurate global emission data for trace halocarbons have played an important role in understanding the chemistry of the atmosphere. In order to calculate emissions, information is needed on the amount produced and the way in which it is used and subsequently emitted to the atmosphere. The first surveys of production and sales were initiated by the manufacturers of chlorofluorocarbons (CFCs) in the mid-1970s as an essential component in the calculation of the atmospheric lifetimes of these gases, which are conserved tracers in the troposphere and reactive in the stratosphere. The producers of methyl chloroform (1,1,1-trichloroethane) followed with data on its annual release, primarily for use in determining average tropospheric abundances of hydroxyl radicals [1]. Methyl chloroform has been the primary compound against which lifetimes of all other halocarbons are calculated. This is made possible by the availability of accurate industry data on this compound.

The industry survey process was extended to include other CFCs and halons, and subsequently to the halocarbon alternatives to CFCs: hydrochlorofluorocarbons and hydrofluorocarbons (HCFCs and HFCs). In the context of the phase-out of methyl chloroform, calculated emissions of certain other shorter-lived compounds, especially tetrachloroethene, are being combined with atmospheric measurements in order to improve understanding of atmospheric processes.

Under the Montreal Protocol, countries are required to report their production and consumption of ozone-depleting substances (ODS) such as CFCs, methyl chloroform, the halons and methyl bromide. The last named is the subject of a separate Chapter in this Handbook [2]. Further information about the provisions of the Montreal Protocol may also be found in the Chapter by Midgley & McCulloch elsewhere in this Handbook [3]. Combination of the complementary data sets from industry and UNEP provides an enhanced tool for continued atmospheric monitoring.

2 Available Sources of Data

The ideal data set for use in assessment models of the atmosphere would have a time history of emissions for each compound with some degree of geographical breakdown, reaching as detailed a level as a $1^\circ \times 1^\circ$ grid for use in 3-dimensional models. Emission data have to be calculated from production and sales data, combined with knowledge of the time history of the release of each compound from its individual applications. The two main data sets considered in this chapter are those from the halocarbon industry and from the United Nations Environment Programme (UNEP).

The strengths and weaknesses of these data sets can be itemised as follows: the industry data (1) are reported for individual compounds, (2) have a long time series, extending back to the start of production for most compounds, (3) sales data are reported in categories in order to calculate emissions, but (4) the degree of geographical breakdown in the data is limited, and (5) global coverage is variable, being virtually total for many compounds but increasingly less so for CFCs in the most recent years. The UNEP data , on the other hand, (1) are reported as combined totals in ODS categories rather than data on individual compounds, (2) extend back only to 1989 plus the base year of 1986 for CFCs and halons, (3) have no breakdown according to use, but (4) they are available on a country-by-country basis, with (5) excellent global coverage of all countries that are Parties to the Montreal Protocol.

Taken together, the complementary data from industry and UNEP make a powerful assessment tool. Thus, judicious combination of the data sets can provide estimates of national releases of individual compounds [4–6]. National emissions can be dispersed, using a proxy model such as a population grid to give emissions on a more detailed scale [7, 8]. Global maps of emissions on a 1° latitude by 1° longitude grid for several of the compounds discussed in this chapter may be found at the following web site address (to be posted in late 1998 or early 1999): *http://groundhog.sprl.umich.edu/geia/rcei.*

Other sources of data for certain compounds include statistics collected nationally by individual countries or regionally, by the European Union, for example. Coverage in time, geography and by compound is by no means complete, and the data are not always easily accessible. However, they can provide a limited test of the estimates derived from the more complete industry and UNEP data sets.

2.1
Industry Data

For over 20 years, the manufacturers of CFCs have organised the collection of audited industrial production and sales data for CFC-11 and CFC-12, and the calculation of emissions from those data [9–11]. The original purpose was to derive atmospheric lifetimes by comparison with atmospheric measurements. In order to provide an estimate of the quantity of material that was not emitted immediately after use but retained within systems and subsequently released, historic production and sales records, right back to the commencement of CFC manufacture in 1931, were extracted by the producers. The survey procedure developed for CFCs 11 and 12 has formed the model for other compounds that have been added subsequently. The list now comprises CFCs 11, 12, 113, 114 and 115 [11, 12], Halons 1211 and 1301 [13], HCFCs 22, 141b, 142b and 124 [11, 14], HFC-134a [11], methyl chloroform [5, 15, 16], and the shorter-lived chlorinated solvents, trichloroethene, tetrachlorethene and methylene chloride [17]. The chemical formula for each CFC, HCFC, HFC and halon mentioned here can be found in the list of Abbreviations.

The surveys were conducted on behalf of the manufacturers by the respective industry group or trade association: for CFCs; this was originally the

Fluorocarbon Program Panel of the Chemical Manufacturers' Association (CMA FPP). In 1990 this activity was taken on by the Alternative Fluorocarbons Environmental Acceptability Study (AFEAS). AFEAS surveys now cover HCFCs and HFCs as well as CFCs. Further information may be found at the AFEAS web site at *http://www.afeas.org*. The European chemical industry council, CEFIC (Conseil Européen des Fédérations de l'Industrie Chimique; web site *http://www.cefic.org*), coordinates the surveys of chlorinated solvents on behalf of the respective trade associations in Europe, Japan and the USA. CEFIC was also responsible for the surveys of halons reported by McCulloch [13].

The manufacturers report the annual production and sales, in pre-defined categories of each compound, to an independent auditor. The published sales totals are defined as sales for dispersive uses only. Any material used as a chemical feedstock is first deducted, because it is chemically transformed and not released to the atmosphere. This is also true for material that is chemically destroyed. Appropriate adjustments are made by the auditor for material that is reported as being recovered and recycled. Opening and closing inventories for each year are reconciled with the preceding and following years, in order to avoid the problem of apparent negative production that sometimes arises with the UNEP data (see Sect. 2.2 below). The published totals are accurate in accountancy terms, having been audited. Quality control of the content is ensured by several stages of review, which – if necessary – include repeated questions to individual reporting companies. The sales totals are used with a suite of release algorithms to generate annual emission totals for each compound (see Sect. 3.2 below).

The reporting producers who are located in Europe, Japan and North America, together with their subsidiaries and production facilities in South America, are able to provide data that covers all of the developed countries of the OECD (Organisation for Economic Cooperation and Development), plus much of the production in the southern hemisphere. This provides essentially complete global coverage for most compounds in most years. It is true for all the surveyed compounds in the early years of their production. The principal exceptions are the most recent years for CFCs 11 and 12, where the fraction surveyed by industry has declined to about a third of the global total (see Sect. 3.1 below). However, even for these CFCs the surveyed fraction remained high (over 80%) for a surprisingly long time, into the 1990s.

The time coverage of the industry surveys is well illustrated in Figs. 1 and 2, which show the annual production reported to AFEAS from the start of production up to 1996, the latest year reported [11]. The starting date ranges from 1931 in the case of CFC-12 to 1991 for HCFC-124, the compound added most recently. The surveys for CFCs 113, 114 and 115 go back only to 1980, so data for the earlier years of production were estimated by back-extrapolation.

The AFEAS companies made a commitment to collect production and sales data on all CFCs, HCFCs and HFCs that are in significant commercial production, i.e. where there are at least 3 AFEAS producers, each with annual production of 1000 tonnes or more for dispersive uses. Commercial confidentiality constrains AFEAS to publish data only if there are more than 2 producers. It is therefore possible that a compound may have had a significant production total

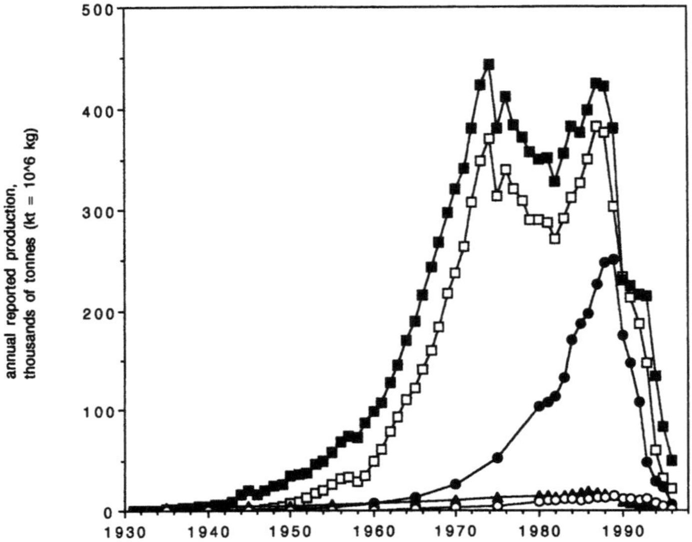

Fig. 1. Annual production reported to industry for all CFCs from the start of production to 1996 [11]. ❑ CFC-11, ■ CFC-12, ● CFC-113, ▲ CFC-114, ○ CFC-115

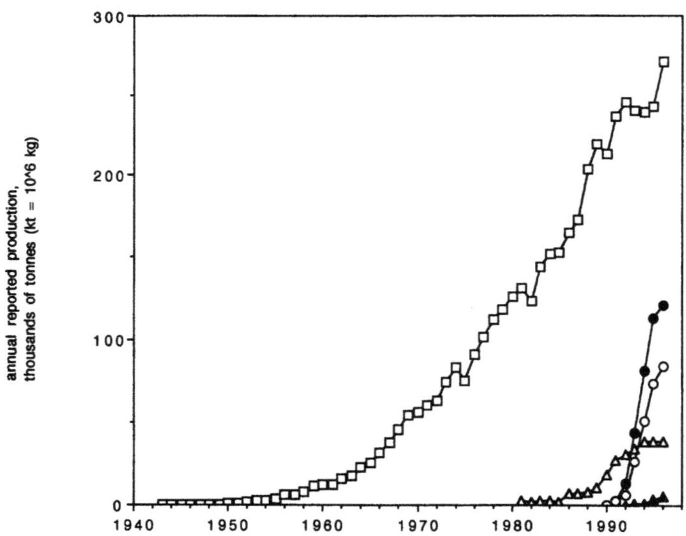

Fig. 2. Annual production reported to industry for all HCFCs and HFC-134a from the start of production to 1996 [11]. ❑ HCFC-22, Δ HCFC-142b, ● HCFC-141b, ▲ HCFC-124, ○ HFC-134a

but – if there were only one or two producers – it would have gone unreported. However, it is likely that the emissions of such a compound would have been detected by atmospheric monitoring and possibly by the difference between AFEAS totals and UNEP totals in the relevant ODS category.

2.2 UNEP Data

The Parties to the Montreal Protocol are required to report their annual production and consumption of ODS to UNEP's Ozone Secretariat [18]. Production is defined under the Protocol as the amount produced minus the amount destroyed minus the amount entirely used as feedstock to manufacture other chemicals, which is essentially identical to the definition used in the industry surveys. Consumption of controlled substances is defined as production plus imports minus exports. Control measures and reporting requirements are prescribed in the Protocol for ODS in specific groups of compounds. The Secretariat arrives at the totals in these groups by multiplying the reported figures for each substance in a group (in metric tonnes) by its ozone depletion potential (ODP) and adding together the ODP-weighted amounts for the group. These aggregated totals are the only data published by UNEP. The groups of controlled substances are as follows:

Montreal Protocol (1987):	Annex A, Group I: CFCs; Annex A, Group II: halons
London Amendment (1990):	Annex B, Group I: other fully halogenated CFCs; Annex B, Group II: carbon tetrachloride; Annex B, Group III: methyl chloroform
Copenhagen Amendment (1992):	Annex C, Group I: HCFCs; Annex C, Group II: HBFCs (hydrobromofluorocarbons); Annex E, Group III: methyl bromide

The quality of the database administered by the Ozone Secretariat is entirely dependent on the quality of the data reports submitted by the Parties. There are many reasons why the database is incomplete for some compounds and/or for some years. Principal among these is the fact that Parties are only required to report on those substances that are controlled by the Protocol and the Amendment(s) that they have ratified. Thus the 162 Parties which had ratified the Montreal Protocol up to 1997 were obliged to report substances in Annex A [18]. The 115 Parties which had also ratified the London Amendment had to report substances in Annexes A and B plus "transitional substances" (HCFCs). Only those 69 Parties which had ratified the Copenhagen Amendment were obliged to report substances in Annexes C and E. There is no requirement for retrospective reporting; data are only required for the years after the entry into force of the Protocol and its Amendments for that particular Party. Moreover developing countries with per capita consumption below the limits prescribed in Article 5 of the Montreal Protocol are entitled to delay control measures, and it is recognised that they may not be able to submit data.

Some confusion has arisen from the UNEP data, because some of the production or consumption figures may be negative. This arises because the figures are reported for each calendar year with no reconciliation of opening and closing inventories, as is done in the industry surveys. The feedstock figure can exceed the production figure for that year if the feedstock material comes from stock carried over from the previous year, giving rise to a negative net production total for the year in question. For similar reasons, the net consumption can also be negative.

An independent review concluded that the quality of data reporting was "at best moderate" [19]. Incomplete reporting by both developing and industrialised country Parties, as well as other deficiencies – such as typographical errors and apparent lack of understanding of the reporting formats – pose major challenges to the accuracy of the global database. Although not all of the problems surrounding its accuracy have been resolved, much progress has recently been achieved by the Ozone Secretariat. Oberthür [19] concluded that the fit between global production and consumption was satisfactory for the major ODS and the quality of the aggregated data was sufficient for analyzing trends. For the few years of data that overlap before the provisions of the Montreal Protocol distorted production (1986 and 1989–1993), there is good agreement between industry data and the subset of UNEP data for the comparable geographic regions.

Methyl chloroform and carbon tetrachloride are the only individual compounds for which global data are published by UNEP, and the database for latter is very deficient. Global totals for individual CFCs may be deduced by combining the aggregated data with the limited amount of information on consumption of individual compounds within economic and geographic regions [20] and with the product distribution in the AFEAS compilations [4]. Before national consumption figures became available from UNEP, a number of surrogates were tried, in order to apportion global CFC totals from the chemical manufacturers on a regional or national scale. Electric power consumption was previously used [21] but it has been found that the relationship between CFC consumption and Gross Domestic Product (GDP) gave results with much lower uncertainties [4]. This proxy relationship can still be used to fill in the gaps in the UNEP data for CFCs and, by analogy, for methyl chloroform [5].

3 Estimation of Global Production and Emission

3.1 Global Production

In the earliest years, all CFC production worldwide was accounted for by the industry surveys, i.e. reported production = global production, and the reported totals with their sales breakdown could be used to derive global emissions. Later, CFCs began to be produced elsewhere in the world, such as in the former Soviet Union, India, China and other developing countries. The manufacturers tried unsuccessfully to find a way to get these production totals included in the audited surveys. Until very recently no reliable data were available from UNEP with which to fill this gap. Therefore, the annual quantity manufactured globally was estimated by applying a factor to the audited data to account for contemporary "unreported" production. Thus, the global production of CFCs 11 and 12 in the 1980s was estimated to be 1.18 times the reported values, based on the reporting manufacturers' informed view that their production represented 85% of the total at that time [22]. The relevant fraction and factor for each year were estimated by combining the reported data with information on global markets. It was assumed that all global production was reported in the industry surveys prior to 1960 and the factors for the intervening years were obtained by interpolation. This simple procedure worked well for CFCs into the 1990s. Sensitivity analysis showed that uncertainties in reported production are translated directly into uncertainties in emissions. However, a ±25% change in non-reported production up to 1991 would have resulted only in a ±1.4% change in the cumulative emissions of CFC-11 and a ±2.5% change in that of CFC-12 because of the relatively small size of this market [23].

Comparisons with information collected to demonstrate compliance with the Montreal Protocol that has recently become available [18] confirm that CFC production reported to AFEAS was of the order of 90% of the global total as late as 1989. During the 1990s, total CFC production – but especially that reported to AFEAS – has fallen steeply due to the controls of the Montreal Protocol. By 1992–3 the industry fraction accounted for around 80% of the total and has continued to fall sharply. The reported component was estimated to represent less than 35% of a much reduced global CFC production total in 1996 [11]. As noted in Sect. 2.2 above, there is a good fit between the data from UNEP and industry for the few years in which they overlap. This lends credence to estimates of global CFC production that must rely primarily on UNEP data in the future.

It is estimated that some 90% of the production of HCFC-22, and all of the production of the other HCFCs and of HFC-134a, is reported to AFEAS [11, 12]. The coverage of methyl chloroform production and use was nearly 100% of the global total [16]. For the three other chlorinated solvents, which are not ODS, production outside the OECD area adds about 10% to the reported production total [17].

3.2 Development of Emission Functions

The industry data set is the only one with the detailed information about uses needed to calculate annual emissions. Release algorithms have been developed based on the pattern of use and the time scale of release from the various applications of each compound. This process is relatively straightforward for those compounds with a single, well-defined use with immediate emissions, such as the use of methyl chloroform or CFC-113 as solvent. It becomes less certain for applications with longer and ill-defined delays between use of the compound under study and its emission to the atmosphere. This is particularly critical for the release of blowing agents from closed-cell foams, where a complex, multi-stage pattern of release takes place over decades. Thus, in general, emission estimates are most accurate for chlorinated solvents and less so for foam blowing agents [24]. There is little or no information on how sales that are not reported to industry are divided among the various uses, so in the first instance it has to be assumed that the use split is identical. Sensitivity analysis has indicated that the error introduced into derived emissions by this assumption would be small, giving a +2.1 to −0.2% change in the cumulative emissions of CFC-11 and a +1.5 to −1.1% change for CFC-12 [23].

Emissions of the material produced in any one year are spread out in time. The most immediate is leakage from the manufacturing process, the "fugitive" emissions that vary from compound to compound but are typically between 2 and 3% of production [10]. Although published sales totals are for dispersive uses only, the auditor makes an allowance for the minor amount lost to the atmosphere during the production, transport and conversion of material to be used as chemical feedstock. The quantity subsequently released after sale depends on the extent to which the compound is contained in the equipment in which it is used and on what happens to it when that equipment is scrapped. For each compound, the end uses can be categorised as either those from which emissions will occur promptly during, or immediately after, use or those where material is retained in a so-called "bank", within systems where emissions may occur over several years or decades. The sales data were, therefore, collected by categories of end use that vary from one compound to another but which gather together sales with similar patterns of potential emissions. The standard procedure will be illustrated by considering CFCs 11 and 12, for which there were originally six categories to report: aerosols, open-cell foams, hermetically sealed refrigeration (domestic refrigerators and freezers), non-hermetic refrigeration (industrial chillers; air-conditioning for both buildings and vehicles), closed-cell foams, and all other uses. The determination of the correct categories and assessment of release delays were made by consulting experts in the relevant markets and applications.

In order to calculate emissions, a prompt-release category was developed by adding sales from other uses with immediate emissions, such as solvents and sterilant gas mixtures, to the aerosol and open-cell foam sales categories. It is assumed that for these uses, release takes place – on average – six months after sale; thus half is released in the year of sale and half the following year. This release category also includes all other sales that have not been attributed to a specific use. These unattributed sales are only a small fraction of the total sales; hence any error in the release estimates arising from lack of their detailed knowledge is negligible [10]. For hermetically sealed refrigeration, leakage takes place predominantly either on charging of the system or on disposal of the appliance, i.e. at the start of the average service life of the system or at its end. In the case of non-hermetic refrigeration, the refrigerant charge life was taken to be 4 ± 2 years, giving a leakage rate of 25% per year. The average service life of the system may be much longer, the equipment being recharged several times during its lifetime.

In the case of foams, there is an initial loss as the foam is blown, followed by losses from handling and trimming the foam plastic. Subsequent losses from closed-cell foams arise from slow leakage of the blowing agent out of the cells of the matrix during the lifetime of the product and – finally – upon disposal. The rate of leakage depends on both the plastic matrix of the foam and the properties of the blowing agent. Ideally, each application should be treated separately, but each CFC blowing agent is used predominantly in one matrix and, to a first approximation, the leakage from this use can be regarded as typical. Thus CFC-12 was used primarily in polyolefin foams, losses from which take place over two years; the same release pattern was assumed to apply to all closed-cell foams blown with CFC-12. For CFC-11, used mainly to blow polyurethane foam, there is an initial loss followed by a much slower seepage of the blowing agent from the foam. This is reflected in the assumption of an initial release of 10% of the quantity used, followed by 4.5% of that quantity each year for the next twenty years [9]. This, or a similar, function has been applied to the closed-cell foam use of each of the other compounds reported. That such a venerable function still represents the best description of the emission from foams is somewhat surprising, given the advances in containment that might have been presumed during 20 years of development, but it has survived reexamination on several occasions [10, 12, 23]. Nevertheless, the function is being examined again, recognising that it represents a significant source of uncertainty for the releases of those compounds that are or were mainly used in blowing closed-cell foams: CFC-11, HCFC-141b and HCFC-142b [11]. For CFC-11, setting the release from foams at the slowest and fastest limits identified would produce an uncertainty of +4.7/−8.4% in the estimated total emissions [23]. Changes in the timing of releases affect cumulative emissions similarly but can have a much larger effect on the annual emission rates calculated, particularly after production has ceased.

For those compounds where most of the releases are prompt, it is obvious that emissions closely follow production, so that 93% of all of the CFC-113 produced by 1990 had been released into the atmosphere by then [12]; nor is there any reason to expect the underlying annual prompt-emission function to have changed since. Similarly, almost 100% of the methyl chloroform sold up to 1993 was estimated to have been released into the atmosphere [5], reinforcing the value of this compound as a proxy for all compounds that react with atmospheric hydroxyl radicals.

4
Comparison of Measured and Calculated Atmospheric Abundances

A long time series of atmospheric measurements exists for the CFCs and methyl chloroform [see for instance 1, 25–28]. These data can be combined with model lifetimes in a chemistry transport model to deduce global emissions. The agreement between these optimally estimated emissions and those estimated independently from industry production and sales data is generally good. Some discrepancies exist, specifically for the most recent years, where the industry data seem to overestimate emissions of CFCs 11 and 12. However, this is perhaps not surprising, in view of the fact that the industry algorithms have not been revised recently to take account of improved use practices, all of which would tend to reduce emissions. For the newer compounds, the halocarbon alternatives to CFCs, there is good agreement between the modelled emissions for HCFC -22 and HFC-134a and those based on data from industry. However the atmospheric abundances of HCFC-141b and HCFC-142b calculated from industry data are smaller than the observed values by factors of about 1.3 and 2 respectively [24]. As mentioned earlier the foam release function for these compounds has greater uncertainties, and has already been targeted by AFEAS for further study.

5 Production, Sales and Emission Data

5.1 Chlorofluorocarbons (CFCs)

A comprehensive data set of production, sales and emissions of CFCs is available for the reasons discussed earlier: the industry record goes back farthest, for many years the industry data had almost complete global coverage and, as controlled species under the Montreal Protocol, data are now available from UNEP that complement and will eventually supersede the industry surveys. The industry survey process and calculation of emissions have not been static, but rather have evolved to take into account both new information and new constraints. As described in Sect. 3.2 above, the original six categories for reporting sales of CFCs 11 and 12 in the industry surveys were: aerosols, open-cell foams, hermetically sealed refrigeration, non-hermetic refrigeration, closed-cell foams, and all other uses. Hermetically sealed refrigeration was only ever a minor application for CFC-12 and, in order to reduce the reporting burden on the companies, the forms were simplified by combining the two refrigeration categories. The previous split between the two types is applied to total refrigeration sales of CFC-12 in order to calculate emissions. Another simplification has been to dispense with the request for CFC sales to be reported according to northern and southern hemispheres. Data in the AFEAS reports from earlier years show that, up to 1992, 95–98% of CFC sales took place in the northern hemisphere.

The original emission functions for CFCs 11 and 12 developed in the 1970s [9] were subsequently amended because of better definition and understanding of certain production and use categories [10]. The uncertainties in the emission estimates and the associated uncertainties in atmospheric lifetimes have been the subject of a sensitivity study [23]. This indicated the need for data on both production and use from countries not covered by the industry surveys, as well as for a better understanding of certain emission categories, in particular, the closed- cell foam application. Non-reported production data are now becoming available from UNEP, but these have no direct categorisation as to use. AFEAS has commissioned a study of the use of halocarbon blowing agents in closed-cell foams [11]. This was primarily aimed at HCFCs, but it may also provide information that will enable further revision of the emissions functions for CFCs.

The production and sales data reported to AFEAS are used to generate an annual report, in which the calculated emissions of CFCs, from reported sales only, are listed by category, i.e., from both types of refrigeration, closed-cell foams and prompt releases. Both annual and cumulative totals are given for the amount produced and released, providing essential information about the size of the "bank". The bank represents material that is held within systems, as yet unreleased, where emissions may occur over years or even decades into the future. It is worth noting that the bank of unreleased material is now far larger than the annual reported production. Moreover, since the early 1990s, annual releases of CFCs 11 and 12 have exceeded the amount produced in the corresponding year because of emissions calculated to come from the bank.

When the survey process was initiated for CFCs 113, 114 and 115 in 1989, it was decided to make a more direct connection between the reporting form and the emissions calculation process by defining the reporting categories directly by the time scale for release. The reporting producers are not constrained by a list of specific applications, but rather attempt to assign all their sales to either of two broad categories based on banking times. This has the effect of dramatically reducing the fraction of total sales that remains unassigned. Thus, uses with short banking times of less than one year (e.g. aerosols, solvents, open-cell foams) are grouped together. The second main category includes all uses with longer banking times (e.g. refrigerants, closed-cell foams, heat transfer and dielectric fluids). Refrigerant applications were characterised by a use lifetime of the order of 12 years for CFCs 113 and 114 and 4 years for CFC-115 [12]. As for CFCs 11 and 12, any other sales that have not been attributed to a specific use are added to the short banking time total for the purpose of the calculation of emissions.

The annual report includes production and sales of CFCs 113, 114 and 115, and their calculated emissions from short and long banking times, on both an annual and cumulative basis. As was the case for CFCs 11 and 12, the bank of unreleased material is now larger than the annual reported production. It is noteworthy that, even for CFC-113, which was sold almost entirely as a solvent and cleaning agent, emissions from the bank now exceed annual sales, because sales for these prompt release applications have declined so rapidly.

Annual data reported to and by AFEAS for production and emission of CFCs 11 and 12 from 1960 to 1996 and for CFCs 113, 114 and 115 from 1970 to 1996 [11] are given in Tables 1, 2, 3 and 4 and are plotted in Figs. 3, 4, 5 and 6. In Fig. 7, the annual sums of CFCs reported to AFEAS are compared with the maximum global CFC production totals allowed under the Montreal Protocol phase-out schedule which began in 1989. The AFEAS data have been summed on an ODP-weighted basis, which means that the annual, reported production of each CFC has first been multiplied by its ozone depletion potential using the ODP values from UNEP [18]. The total ODP-weighted, reported CFC production fell by more than 90% from the peak year, 1988, to 1996.

In order to estimate world production of CFCs 11, 12 and 113, for the years prior to 1990, the annual totals reported to industry were multiplied by the relevant factor, in order to take contemporary, non-reported production into account, as described in Sect. 3.1. The resultant global values for 1960–1989 are identical to those given by Fisher et al. [22]; the data for earlier years may also be found in that work. For production after 1990, the method of calculation has been changed. The industry data are now supplemented by the necessary

Table 1. Annual production and calculated emissions of CFC-11, 1960–96 (in kt per year). From AFEAS [11] and global estimates

Year	AFEAS Reporting Companies only		Global Estimates	
	Production	Emission	Production	Emission
1960	49.7	40.5	50.1	40.9
1961	60.5	52.1	61.0	52.6
1962	78.1	65.4	78.7	65.9
1963	93.3	80.0	94.0	80.6
1964	111.1	95.0	111.9	95.7
1965	122.8	108.3	123.8	109.1
1966	141.0	121.3	142.2	122.3
1967	159.8	137.6	161.1	138.8
1968	183.1	156.8	184.7	158.1
1969	217.3	181.9	219.9	183.8
1970	238.1	206.6	241.0	209.0
1971	263.2	226.9	266.6	229.8
1972	306.9	255.8	311.1	259.2
1973	349.1	292.4	354.3	296.6
1974	369.7	321.4	377.5	327.1
1975	314.1	310.9	323.4	318.4
1976	339.8	316.7	350.5	325.9
1977	320.5	303.9	332.8	314.0
1978	308.9	283.6	323.0	294.9
1979	289.5	263.7	305.8	276.1
1980	289.6	250.8	308.4	264.6
1981	286.9	248.2	308.6	263.8
1982	271.4	239.5	296.3	257.2
1983	291.7	252.8	320.4	273.8
1984	312.4	271.1	345.4	295.6
1985	326.8	280.8	364.8	308.6
1986	350.1	295.1	393.8	326.8
1987	382.1	310.6	431.2	345.7
1988	376.0	314.5	431.2	353.5
1989	302.5	265.2	364.6	304.7
1990	232.9	216.1	262.2	244.0
1991	213.5	188.3	242.8	211.0
1992	186.4	171.1	215.7	192.9
1993	147.1	157.9	176.4	179.8
1994	60.2	137.4	89.5	160.8
1995	32.7	123.8	62.0	149.0
1996	22.1	118.2	57.7	150.9

numbers extracted from the UNEP database for production in developing countries [18, 24]. Emissions are calculated by applying the AFEAS emission algorithms to the new production totals. Analysis of the limited data available from UNEP on individual compounds has shown that production totals reported to AFEAS for CFCs 114 and 115 account for essentially all of the global production [4]; hence, in this work the global totals are equated to the AFEAS totals. The

Table 2. Annual production and calculated emissions of CFC-12, 1960–96 (in kt per year). From AFEAS [11] and global estimates

Year	AFEAS Reporting Companies only		Global Estimates	
	Production	Emission	Production	Emission
1960	99.4	89.1	102.4	91.6
1961	108.5	99.7	111.9	102.7
1962	128.1	114.5	132.2	118.0
1963	146.4	133.9	151.2	138.2
1964	170.1	155.5	175.8	160.6
1965	190.1	175.4	196.8	181.3
1966	216.2	195.1	224.1	201.8
1967	242.8	219.9	252.1	228.0
1968	267.5	246.5	278.8	256.2
1969	297.3	274.3	311.4	286.2
1970	321.1	299.9	336.8	313.8
1971	341.6	321.8	360.5	338.0
1972	379.9	349.9	401.7	368.9
1973	423.3	387.3	447.4	408.7
1974	442.8	418.6	473.5	444.3
1975	381.0	404.1	418.5	435.5
1976	410.7	390.4	450.3	425.2
1977	382.8	371.2	424.5	406.1
1978	372.1	341.3	416.0	376.7
1979	357.2	337.5	403.4	375.7
1980	350.2	332.5	399.0	373.3
1981	351.3	340.7	402.7	385.3
1982	328.0	337.4	382.2	385.3
1983	355.3	343.3	412.4	394.4
1984	382.1	359.4	442.3	413.6
1985	376.3	368.4	439.7	426.1
1986	398.4	376.5	465.2	437.5
1987	424.7	386.5	499.8	451
1988	421	392.8	505.5	462.7
1989	379.8	364.7	474.8	436.9
1990	231.0	310.5	305.0	378.8
1991	224.8	271.6	298.8	335.7
1992	216.2	255.3	290.2	319.9
1993	214.7	237.8	288.7	302.8
1994	133.6	211.5	207.6	277.3
1995	82.8	188.6	156.9	257.8
1996	48.9	166.1	138.4	242.0

global values for production and emission of CFCs are displayed with the AFEAS values in the Tables and Figures mentioned above.

Global production in 1996 was less than 15% of that in the year of peak production for CFC-11 and less than 30% of the peak amount for CFC-12. As can be seen in Figs. 3 and 4, the reductions in calculated emissions lag behind those in reported production, because emissions now mostly come from the bank. Global

Table 3. Annual production and calculated emissions of CFC-113, 1970–96 (in kt per year). From AFEAS [11] and global estimates

Year	AFEAS Reporting Companies only		Global Estimates	
	Production	Emission	Production	Emission
1970	26.6	25.0	29.8	28.0
1971	30.5	28.6	34.2	32.1
1972	35.0	32.9	39.3	36.9
1973	40.0	37.6	44.9	42.2
1974	45.9	43.1	51.5	48.4
1975	52.6	49.4	59.0	55.5
1976	60.2	56.6	67.5	63.5
1977	69.0	64.9	77.4	72.8
1978	79.0	74.3	88.6	83.4
1979	90.5	85.1	101.5	95.5
1980	103.7	97.5	116.3	109.4
1981	108.5	106.4	121.8	119.4
1982	113.0	111.1	126.8	124.6
1983	132.7	123.3	148.9	138.3
1984	171.1	152.5	192.0	171.1
1985	187.0	179.8	209.8	201.7
1986	196.6	193.1	220.6	216.6
1987	225.8	213.1	248.1	236.4
1988	247.4	238.7	268.1	260.3
1989	251.3	251.5	270.3	271.6
1990	174.8	214.7	195.7	233.8
1991	147.6	163.1	168.5	181.5
1992	107.5	128.3	128.4	147.5
1993	48.0	78.0	68.9	80.5
1994	29.5	39.9	50.4	58.7
1995	23.3	27.6	44.2	50.8
1996	6.0	15.7	31.1	33.7

emissions of CFC-11 and CFC-12 in 1996 were calculated to be about 40% and 50%, respectively, of those in the peak year. For CFC-113, however, there is little lag apparent, because it was used in solvent cleaning with a short release delay, and both global production and emissions had fallen to 15% of the peak value by 1996 (Fig. 5). In contrast, CFC-115 was used mostly in a refrigeration application and the lag is greater. Global production had fallen to less than 15% of the peak value by 1996, but emissions were calculated still to be 75% of the peak value.

As discussed in section 2 above, global production, sales and emission totals can be distributed among countries on the basis of their economic activity, as indicated by the national GDP [4]. Emissions from the national distributions have subsequently been mapped onto a 1° latitude by 1° longitude grid, using the distribution of population within each country reported by Logan. This results in global maps of emissions, where each grid square represents a region of about 100 km by 100 km [29]. Maps for CFCs 11 and 12 have been produced for the

Table 4. Annual production and calculated emissions of CFC-114 and CFC-115, 1970–96 (in kt per year). AFEAS reporting companies = global estimates [11]

Year	CFC-114		CFC-115	
	Production	Emission	Production	Emission
1970	10.2	9.7	2.3	1.3
1971	10.6	10.1	2.6	1.6
1972	11.1	10.5	3.1	1.9
1973	11.5	10.9	3.6	2.2
1974	11.9	11.3	4.2	2.5
1975	12.4	11.7	4.8	3.0
1976	12.9	12.2	5.6	3.5
1977	13.4	12.7	6.5	4.0
1978	13.9	13.2	7.6	4.7
1979	14.4	13.7	8.9	5.4
1980	15.0	14.2	9.3	6.3
1981	14.0	14.2	10.0	7.2
1982	13.6	13.7	10.4	8.1
1983	14.8	14.1	11.6	8.9
1984	15.6	15.1	11.2	9.6
1985	17.1	16.2	10.0	10.1
1986	19.1	18.0	11.8	10.6
1987	17.1	18.2	12.8	11.0
1988	16.5	16.2	13.6	11.4
1989	15.0	14.5	14.2	11.9
1990	8.3	10.3	11.3	12.2
1991	6.7	6.3	12.3	12.6
1992	4.7	5.2	10.7	12.6
1993	4.6	4.6	11.4	12.6
1994	3.2	4.0	6.8	11.8
1995	3.1	3.1	3.7	10.9
1996	0.7	2.4	2.0	9.5

Global Emissions Inventory Activity (GEIA) and may be found on the web site at *http://groundhog.sprl.umich.edu/geia.*

5.2 Halons

Halons which are used as specialised fire extinguishing agents are regulated as ODS under the Montreal Protocol. In 1990, CEFIC conducted a survey on behalf of the halon producers within the OECD, in which annual production totals of Halons 1211 and 1301 were reported [13]. Data prior to 1980 for Halon-1211 and prior to 1972 for Halon-1301 had to be aggregated before publication, because of the small number of producers. Estimates of annual production for the years from the start of production in the early 1960s to these dates were derived by back-extrapolation of the growth curves. The reporting producers accounted for all production within the OECD. Known production elsewhere prior to 1988

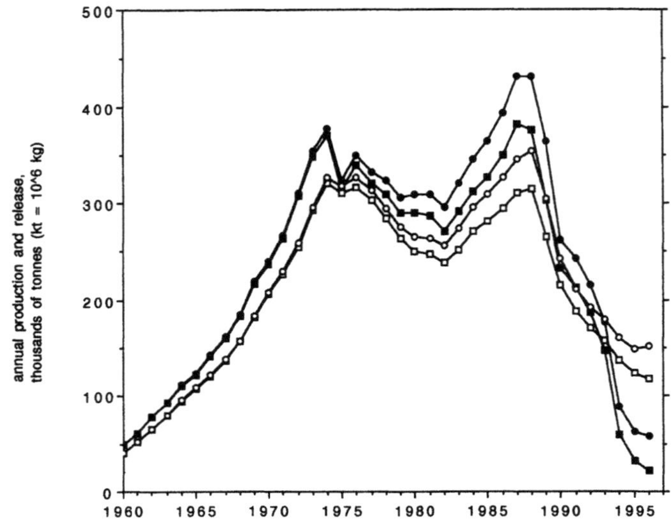

Fig. 3. Annual production and emission of CFC-11 (from AFEAS [11] and global estimates), 1960–96. ■ Production reported to AFEAS, ❑ calculated emission, reporting companies only, ● estimated global production, ❍ calculated global emission

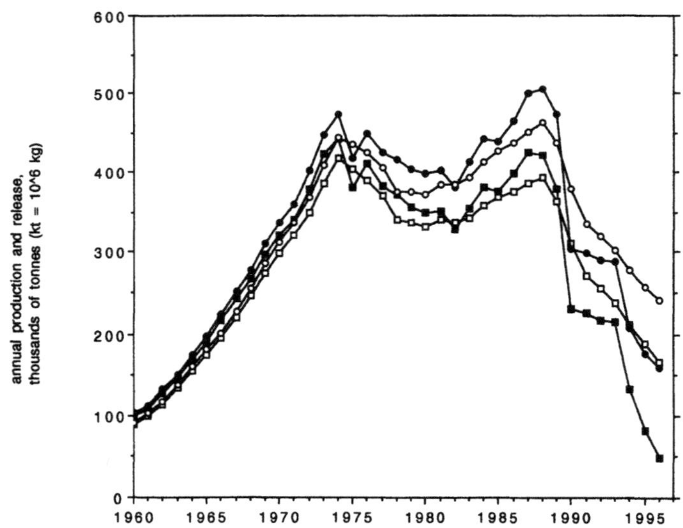

Fig. 4. Annual production and emission of CFC-12 (from AFEAS [11] and global estimates), 1960 – 96. ■ Production reported to AFEAS, ❑ calculated emission, reporting companies only, ● estimated global production, ❍ calculated global emission

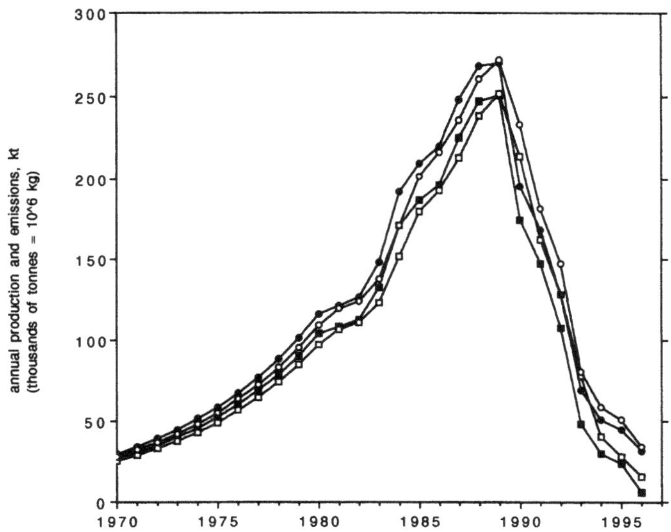

Fig. 5. Annual production and emission of CFC-113 (from AFEAS [11] and global estimates), 1970–96. ■ Production reported to AFEAS, ❑ calculated emission, reporting companies only, ● estimated global production, ❍ calculated global emission

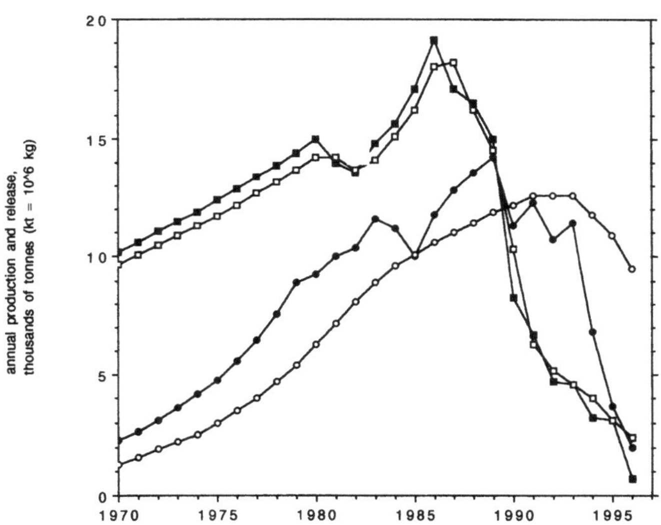

Fig. 6. Annual production and emission of CFC-114 and CFC-115 (AFEAS reporting companies = global estimates), 1970–96 [11]. ■ production of CFC-114, ❑ calculated emission of CFC-114, ● production of CFC-115, ❍ calculated global emission of CFC-115

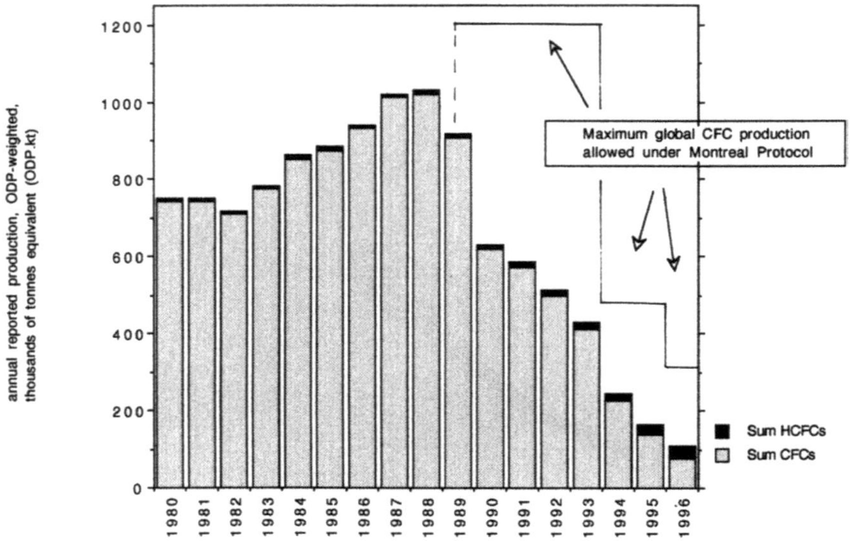

Fig. 7. ODP-weighted fluorocarbon production, 1980–96 [11]. Maximum global production allowed under the Montreal Protocol from [18]. *Light shaded blocks* ODP-weighted sum of all CFCs reported to AFEAS, *dark shaded blocks* ODP-weighted sum of all HCFCs reported to AFEAS. Note HFCs contain no chlorine and therefore their ODP is zero

consisted of a limited amount of Halons 1211 and 1301 produced in China. Production in Russia and Eastern Europe was predominantly Halon-2402. Estimates of global production were made by the addition of a small amount to the reported OECD totals [13]. Data have now become available from UNEP which substantially confirm these earlier estimates and can be used to provide a better assessment of global production from 1989 onwards [18, 30]. Global production values for 1970–1996 are shown in Table 5 and Fig. 8.

The global production of Halons 1211 and 1301 peaked in 1988. Production ceased in the developed world in 1994 but permitted production is known to occur in developing countries (China, India, Korea) [18]. However large quantities of these gases are stored in equipment and global emissions come predominantly from this bank. Using a combination of audited industry data and national data reported to UNEP, estimates have been made of the annual emissions of Halons 1211 and 1301; these are also shown in Table 5 and plotted in Fig. 8 [13, 24, 30]. To illustrate the size of the bank, note that – although production of Halon-1211 at its peak in 1988 was 21,000 tonnes and emissions were about 13,000 tonnes, – the bank was 51,000 tonnes and had grown by 1994–6 to an estimated 60,000 tonnes. The equivalent figures for Halon-1301 were 1988 production: 13,000 tonnes, 1988 emissions: 4000 tonnes, 1988 bank: 60,000 tonnes, 1994–6 bank: about 70,000 tonnes [30].

Halon-2402 was used almost exclusively in Eastern Europe, including Russia, and very little Halon 1211 or 1301 was manufactured there. It can be deduced

Table 5. Annual global production and emission estimates of Halons 1211 and 1301, 1963–1996 (in kt per year) [24]

Year	Halon-1211		Halon-1301	
	Production	Emission	Production	Emission
1963	0.1	0.0		
1964	0.1	0.1		
1965	0.2	0.1		
1966	0.3	0.2		
1967	0.5	0.3	0.1	0.0
1968	0.7	0.5	0.1	0.0
1969	0.9	0.6	0.1	0.0
1970	1.3	0.8	0.2	0.1
1971	1.7	1.1	0.6	0.2
1972	2.2	1.5	0.8	0.3
1973	2.8	1.8	1.3	0.5
1974	3.3	2.2	1.5	0.5
1975	3.8	2.5	2.0	0.7
1976	4.4	2.9	3.2	1.1
1977	5.0	3.3	3.5	1.3
1978	5.7	3.7	4.0	1.4
1979	6.3	4.1	4.7	1.7
1980	6.9	4.6	4.9	1.8
1981	6.7	4.6	5.7	3.0
1982	7.5	5.2	7.6	3.9
1983	9.3	6.8	7.4	3.8
1984	11.4	8.5	8.7	4.5
1985	14.3	10.8	9.8	5.1
1986	15.4	10.0	11.1	3.7
1987	18.3	11.6	11.6	3.8
1988	20.2	12.7	12.6	4.1
1989	16.3	10.3	11.2	3.7
1990	15.1	10.3	9.2	3.4
1991	13.2	10.4	8.1	3.9
1992	11.0	10.2	5.8	3.9
1993	9.1	9.6	3.9	3.4
1994	5.3	8.9	0.9	3.2
1995	10.7	7.8	0.9	3.1
1996	10.7	7.4	0.9	3.0

from the UNEP data that production of Halon-2402 was about 4000 tonnes in 1986, falling to less than 10% of this total in 1994–5 [18].

5.3
Hydrochlorofluorocarbons and Hydrofluorocarbons (HCFCs and HFCs)

Data collection for HCFC-22 was initiated in 1990 mainly in order to provide another candidate compound for the assessment of hydroxyl radical (OH) concentrations in the atmosphere, because methyl chloroform, the primary

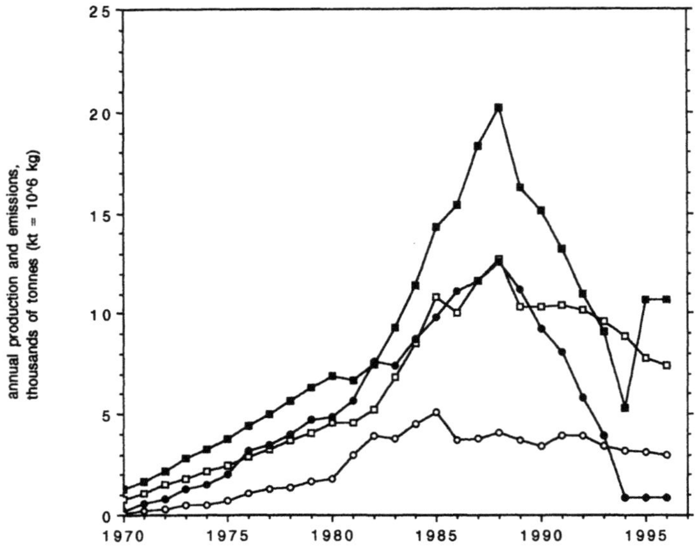

Fig. 8. Annual global production and emissions of Halons 1211 and 1301 [24]. ■ Production of H-1211, ❑ calculated emission of H-1211, ● production of H-1301, ❍ calculated emission of H-1301

compound for this study, was being phased out [14]. As the most widely used HCFC at that time, particularly as a substitute for CFCs in refrigeration systems, it was studied as a surrogate for the family of HCFCs that were beginning to be produced in increasing amounts as substitutes for CFCs in various applications. The production and sales survey was designed following the model of the earlier CFC and methyl chloroform surveys. One difference is that a significant fraction of HCFC-22 production is used as a chemical feedstock in the production of fluoropolymers. The auditor subtracts this amount from the total production because it is consumed within the process and, therefore, is not emitted to the atmosphere. The resultant, published net production total represents sales into dispersive or emissive uses. The amount used as a chemical feedstock is not published for reasons of commercial confidentiality, but information provided to an earlier WMO Science Assessment indicates that it was about 35% of the total produced [31]. Fugitive emissions from chemical processing, as well as those from the manufacture and packing of the original HCFC, are taken into account via estimates derived from process information provided by the manufacturers [14].

Annual production totals only for HCFC-22 (for the years 1970–1979) were reported , as were both production and sales data after 1980. Reported sales are grouped in three categories, according to the time delay between sale and release. Over 90% of sales for dispersive uses are for refrigeration and air-conditioning applications, with a range of service lifetimes between 3 and 20 years. A composite emission function was developed to describe this complex release pattern

Table 6. Annual production and calculated emissions of HCFC-22, 1970–96 (in kt per year). From AFEAS [11] and global estimates

Year	AFEAS Reporting Companies only		Global Estimates	
	Production	Emission	Production	Emission
1970	56.1	41.4	61.7	45.5
1971	60.6	45.8	66.7	50.4
1972	63.2	50.1	69.5	55.1
1973	74.3	57.5	81.7	63.2
1974	83.4	65.0	91.7	71.5
1975	75.0	65.7	82.5	72.3
1976	90.7	74.0	99.8	81.4
1977	101.4	82.4	111.5	90.6
1978	111.7	90.1	122.9	99.1
1979	117.9	96.7	129.7	106.4
1980	126.3	104.3	138.9	114.7
1981	130.8	110.3	143.9	121.3
1982	123.6	111.9	136.0	123.1
1983	143.9	122.8	158.3	135.1
1984	152.4	131.8	167.6	145.0
1985	153.4	137.2	168.7	150.9
1986	165.0	145.6	181.5	160.2
1987	173.3	153.7	190.6	169.1
1988	203.5	171.4	223.9	188.5
1989	219.5	188.7	241.5	207.6
1990	213.7	195.2	235.1	214.7
1991	236.8	204.3	260.5	224.7
1992	245.7	214.8	270.3	236.3
1993	240.6	212.7	264.7	234.0
1994	239.4	218.9	263.3	240.8
1995	243.5	224.7	267.8	247.2
1996	271.2	242.0	298.4	266.2

[14]. The annual sales total is also broken down into three regions: 0–30 °N, 30–90 °N and the southern hemisphere. This degree of geographical breakdown, greater than that required for the CFCs, is important for model studies. This is because a single, global figure for annual emissions of HCFC-22 would not be sufficient to constrain calculations of the OH abundance, which varies greatly in time and space. In practice, approximately 90% of all HCFC-22 produced for emissive uses has been sold in the industrial north, i.e. in the region 30–90 °N. Annual data reported to and by AFEAS on the production and emission of HCFC-22 from 1970 to 1996 [11] are given in Table 6 and plotted in Fig. 9.

Production of HCFCs is known to take place in China, India, Korea, and the Russian Federation [18]. Although UNEP publishes national data on HCFC production and consumption, these are only reported as aggregated totals in ODP-tonnes for all of the HCFCs together. Thus even if the data were complete, which they are not for all years, it would be impossible to extract information from them on the use of HCFC-22 alone. Comparison of the data from AFEAS and

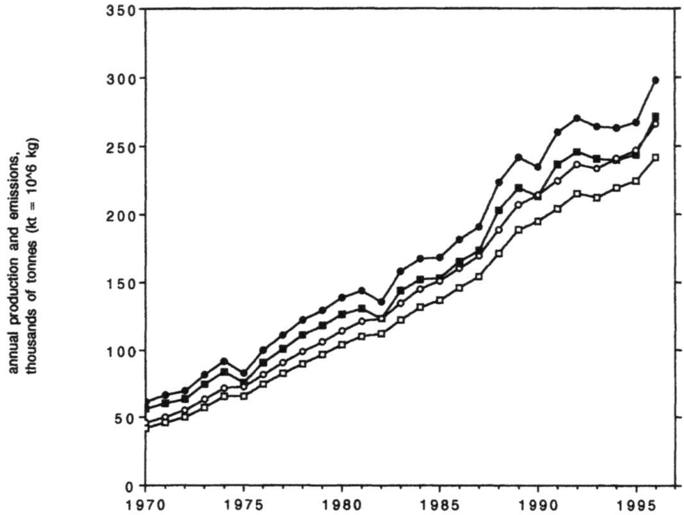

Fig. 9. Annual production and emission of HCFC-22 (from AFEAS [11] and global estimates), 1970–96. ■ Production reported to AFEAS, ❑ calculated emission, reporting companies only, ● estimated global production, ○ calculated global emission

UNEP for 1989 confirms the manufacturers' earlier estimate that production reported to AFEAS, primarily HCFC-22 in that year, amounted to 90% of the global total. Moreover, it was estimated that the AFEAS reports account for well over 90% of the world total production of HCFC-22 for dispersive uses. Global production of HCFC-22 has been calculated by applying a simple, 10% factor to the AFEAS totals. Sales and emission patterns are unlikely to differ globally, so global emission estimates for HCFC-22 can be derived using the AFEAS algorithms and are also displayed in Table 6 and Fig. 9. Allowing for the uncertainties in "non-reported" production and use, there is reasonable agreement between atmospheric abundances derived from the calculated global emissions and those observed [27, 28].

The distribution of CFC-12 emissions among countries [4] has been used in calculations as a surrogate for the distribution of HCFC-22 emissions; this is a valid procedure, given that both are refrigerants [6]. There had indeed been a gradual change to the use of HCFC-22 in place of CFC-12 that predates the regulations under the Montreal Protocol. In a similar procedure to that described above for CFCs, the population density reported by Li [32] has been used to distribute these national emissions further onto a 1° latitude by 1° longitude grid [6, 7]. The resultant map will be posted before the end of 1998 at the Reactive Chlorine Emissions Inventory (RCEI) website: *http://groundhog.sprl.umich.edu/geia/rcei.*

The companies reporting to AFEAS made a commitment to collect production and sales data on all CFCs, HCFCs and HFCs that are in significant, commercial production and to publish them, within the constraints of commercial confidentiality. To date, this has added surveys on HCFC-142b (1993), HCFC-141b (1995) and HCFC-124 (1997). The only HFC which has met both criteria so far has been HFC-134a (1995), the most widely used HFC [11]. In all cases but one, production and sales records were extracted right back to the start of production, in order to estimate atmospheric abundances accurately. The exception was HCFC-142b, which had a longer history of small scale commercialisation during the 1970s as well as manufacture for use as a chemical feedstock. Annual data on HCFC-142b that have been provided go back to 1981, but records for earlier years were not available. In the period 1981–1985, however, production averaged only 2000 tonnes per year for dispersive uses.

Similar survey forms have been adopted for all four of the "new" AFEAS compounds, requiring information about sales in five categories: all refrigeration and air-conditioning applications, closed-cell foams, medical aerosols (metered dose inhalers), all solvent applications, and a final category for all uses with short-term emissions. This fifth category includes use in all other types of aerosol, as sterilants or as blowing agents for open-cell foams. HCFC-142b and HCFC-141b are mainly used as blowing agents for closed-cell foams (over 90% and 85% of total sales, respectively), and solvents account for a further 10% of sales of HCFC-141b. HFC-134a is used mainly in refrigeration and air-conditioning applications (over 85% of total sales), whereas some 60% of HCFC-124 sales are for these applications and the other 40% are accounted for by uses with a short release delay, such as sterilants. In each case, the emission algorithms based on CFC-use practices have been adapted for the replacement of CFCs by HCFCs in specific applications.

HCFCs are regulated as transitional substances under the Copenhagen Amendments to the Montreal Protocol [3]. National data are collected by UNEP from the Parties, but these are published as the ODP-weighted sum of all HCFCs [18]. Thus it is not possible to get information on individual HCFCs from the UNEP data. It is believed that the manufacturers reporting to AFEAS account for virtually all of the global production of HCFCs 142b, 141b and 124 and HFC-134a [11]. Thus the reported production data and the emission data derived from them, for these compounds, displayed in Table 7 and Figs. 10 and 11, can be taken to represent global production and emissions.

Emissions calculated from the industry sales and use data are in accordance with the current, measured global abundances of HCFC-22 and HFC-134a. For HCFCs 142b and 141b, the industry data appear to underestimate the measured abundances by factors of 2 and 1.3, respectively [24]. It has long been acknowledged that the algorithm for the release of blowing agents from closed-cell foams was less certain than those for other applications because of its relative complexity [12]. Thus, emission estimates for compounds such as HCFCs 142b and 141b have greater uncertainties, principally because of the difficulties in defining and quantifying release from foams. As mentioned earlier in this Chapter, this application is being studied in order to improve the algorithm and to base it on up-to-date information specific to HCFCs [11].

In Fig. 7, the annual sums, on an ODP-weighted basis, of all HCFCs reported to AFEAS have been added to the CFC sums. This demonstrates that, although the production and use of HCFCs has grown considerably since 1990, their

Table 7. Annual reported production and calculated emissions of HCFC-142b, 1981–96; HCFC-141b, HCFC-124 and HFC-134a, 1990–96 (AFEAS reporting companies=global estimates) (in kt per year) [11]

Year	HCFC-142b		HCFC-141b		HCFC-124		HFC-134a	
	Production	Emission	Production	Emission	Production	Emission	Production	Emission
1981	2.6	0.6						
1982	1.9	0.5						
1983	2.2	0.5						
1984	2.4	0.6						
1985	1.4	0.4						
1986	7.1	1.6						
1987	6.9	1.7						
1988	7.8	2.0						
1989	10.3	2.9						
1990	18.8	5.8	0.1	0.0			0.2	0.1
1991	27.2	8.4	1.5	0.4	0.02	0.0	2.2	0.3
1992	30.7	10.8	13.3	3.9	0.1	0.1	6.4	0.9
1993	33.7	10.2	43.3	13.1	0.5	0.3	26.5	3.7
1994	38.4	10.7	81.2	24.8	0.9	0.4	50.4	9.4
1995	38.7	12.0	113.2	36.6	3.1	1.8	73.8	20.3
1996	38.1	11.7	121.1	39.5	4.9	3.5	83.7	32.6

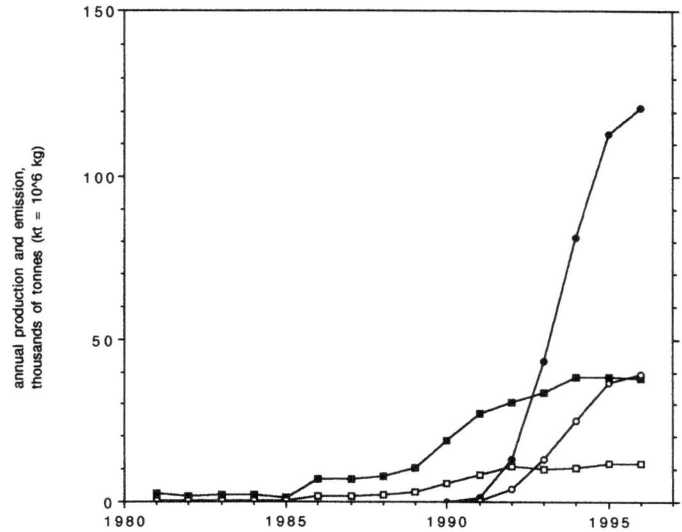

Fig. 10. Annual production and emissions of HCFC-141b and HCFC-142b (AFEAS reporting companies = global estimates), 1981–96 [11]. ■ Production of HCFC-142b, ❑ calculated emission of HCFC-142b, ● production of HCFC-141b, ❍ calculated emission of HCFC-141b

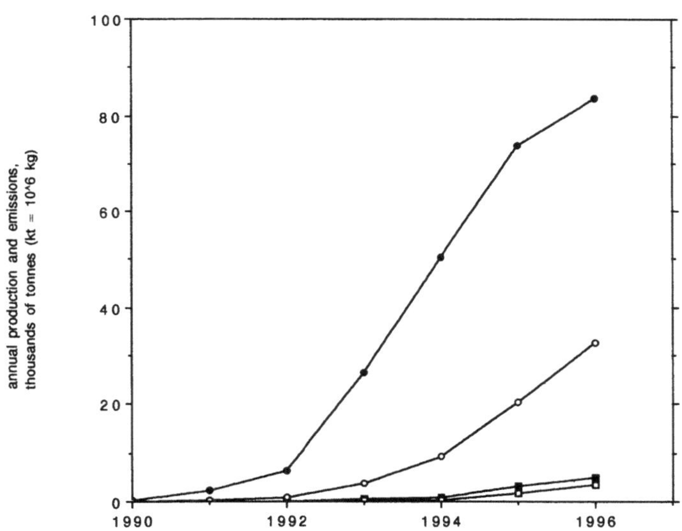

Fig. 11. Annual production and emissions of HCFC-124 and HFC-134a (AFEAS reporting companies = global estimates), 1990–96 [11]. ■ Production of HCFC-124, ❑ calculated emission of HCFC-124, ● production of HFC-134a, ❍ calculated emission of HFC-134a

contribution to ozone depletion potential is insignificant compared to that of the CFCs that they have replaced.

HCFC-123 does not meet the AFEAS reporting criterion that three significant producers must submit data to the auditor before any numbers can be released. However, it was included in a recent analysis of sales of all HCFCs in the European Union (EU) [33]. Sales of HCFC-123 in the EU in 1995, principally in refrigeration, were estimated to total 4500 tonnes, a ten-fold increase since 1990, but still an order of magnitude less than sales of HCFC-22.

HFC-23, produced as a byproduct during the manufacture of HCFC-22, has been used until recently as a starting material in the production of Halon-1301. However, such use has dropped to near zero as a result of the phase-out of halon production in developed countries under the Montreal Protocol [24]. Some limited use continues in plasma etching processes, as a low temperature refrigerant and as a fire suppressant. Its atmospheric abundance is consistently about 10% of that of HCFC-22 [34].

5.4
Chlorinated Solvents

This section describes the data available on four chlorinated hydrocarbons, all used principally as solvents. Of these, only methyl chloroform is regulated as an ODS under the Montreal Protocol. The other three chlorinated solvents: trichloroethene, tetrachlorethene and methylene chloride, are much shorter-lived -with atmospheric lifetimes of one week, 3–4 months and 5–6 months, respectively – compared to the 4.8 years of methyl chloroform, and thus do not cause concern for stratospheric ozone depletion [24].

Audited data on the annual release of methyl chloroform have been provided by industry for use in determining average tropospheric abundances of hydroxyl radicals [1, 15]. Methyl chloroform has been the primary compound against which the atmospheric lifetimes of all other halocarbons were calculated. Apart from use as a chemical feedstock, the applications of methyl chloroform are those such as solvent cleaning that, result in rapid emissions. Unless specific steps are taken to capture and destroy used material, it is ultimately released into the atmosphere, so that emissions are equivalent to the amount sold, with release taking place – on average – within six months of sale [15]. There was some evidence of stock building in certain markets, in anticipation of regulation, that introduced a somewhat greater delay between production and release in later years [5]. This has been accounted for in the estimates of emissions that were developed based on the industry data. The annual surveys were organised by CEFIC on behalf of the world's methyl chloroform producers and cover production in the OECD from 1970 to 1995 and sales throughout the world [35]. Sales from 1980 onwards were broken down into six discrete geographical regions and into three use categories, although in practice virtually all sales fell into the short release category. The methyl chloroform manufacturers estimated that their reported data have almost complete global coverage and this has subsequently been confirmed by comparison with data reported to UNEP to monitor compliance with the Montreal Protocol [5, 15, 18].

Table 8. Annual reported sales by region and calculated global emissions of methyl chloroform, 1980–96 (in kt per year) [5, 16, 35]

Year	Sales by region (reporting companies only)						Calculated Global Emissions
	90–0°S	0–30°N	30–90°N				
			N. America	Europe	Far East	Other	
1980	24	4	276	156	88	2	538
1981	21	6	262	161	96	2	549
1982	21	6	228	160	96	2	523
1983	20	8	235	164	113	2	536
1984	22	9	267	166	132	4	585
1985	22	8	246	167	141	3	593
1986	27	16	249	166	148	3	602
1987	25	16	251	163	167	5	666
1988	27	18	271	160	194	9	666
1989	26	20	260	165	211	11	691
1990	20	25	274	155	245	8	718
1991	21	35	204	136	201	8	635
1992	19	55	233	118	193	8	593
1993	17	22	107	79	70	7	380
1994	5	31	67	87	64	1	283
1995	4	21	50	91	52	2	234
1996	0	0	0	0	0	0	65

The audited sales within each region and the global emission estimates derived from these for 1980–1996 are given in Table 8. For the purposes of this calculation, it has been assumed that the sales totals reported by industry and hence the calculated emission totals are effectively global. Production totals for earlier years may be found in the references previously cited [5, 15]. Sales within the six regions have been dispersed onto a national scale using CFC-113 sales as a proxy; this is reasonable, as both were used principally as solvents in similar industrial applications [5]. Emissions from this national distribution have been mapped onto a 1° latitude by 1° longitude grid, using the population density reported by Li [29, 32]. This map will also be available at the RCEI web site at the address given previously.

As a result of controls introduced under the Montreal Protocol, the production and consumption of methyl chloroform have now been phased out in developed countries, apart from production for feedstock use and any material supplied to meet the needs of developing countries. The phase-out schedule for those countries operating under Article 5 of the Montreal Protocol requires a freeze in production and consumption in 2003 followed by a phase-out in 2015 [18]. As can be seen in Fig. 12, industrial sales of methyl chloroform have declined rapidly in recent years in all regions; sales in 1995 were less than one-third of the amount reported for 1990, the year of peak production. No new production was reported in 1996 on sales for emissive uses in the developed world [16]. There, any production was solely for use as chemical feedstock. Also shown in

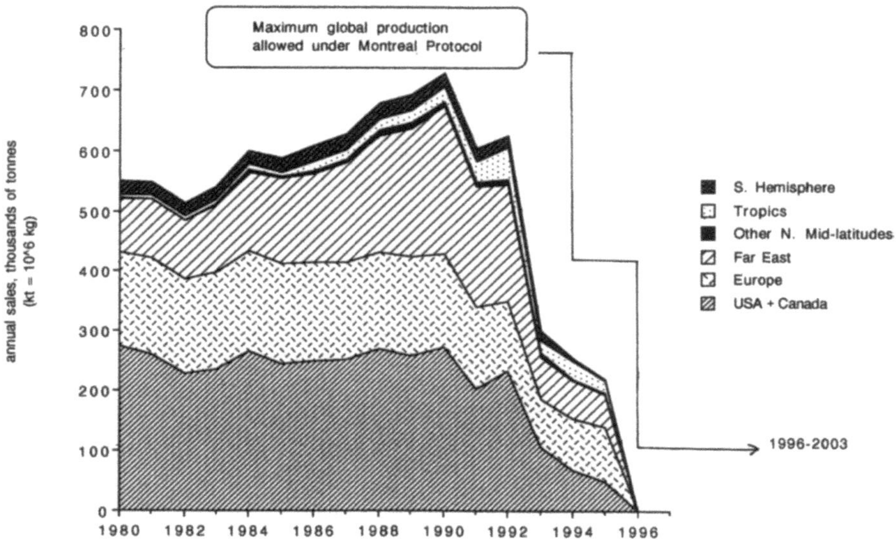

Fig. 12. Regional sales of methyl chloroform reported to CEFIC, 1980–1996 [5, 16, 35]. Maximum global production allowed under the Montreal Protocol from [18]

Fig. 12 are the maximum global production totals allowed under the Montreal Protocol from 1993 onwards. Total consumption of methyl chloroform in developing countries operating under Article 5 of the Montreal Protocol was reported to be less than 24,000 tonnes in 1996 [36]. It has been estimated that these requirements were fulfilled mostly by material exported from developed countries and that the production capacity in developing countries lies well below this figure [5]. Assuming that the use-patterns of recent years continue to apply, it has been estimated that emissions in 1996 would be 65,000 tonnes, dropping to 10,000 tonnes in 1997 and – with no new production to sustain continued emission – would tail off over the next 5 years [16]. If material had been stockpiled for future use ahead of the phase-out, it is possible that emissions for the past few years have been overestimated and future emissions underestimated, but the effect on atmospheric concentrations will be small.

The decrease in production and sales of methyl chloroform was quickly followed by a decline in atmospheric emissions, because most applications result in its rapid release to the atmosphere soon after use. The comprehensive time-series of measurements available from networks of monitoring stations distributed world wide has clearly demonstrated that there has been a worldwide decline in atmospheric abundance, mirroring the decline in its sales and derived emissions [1, 27]. Because of the short global lifetime of this compound, changes in its abundance have a prompt impact on the level of atmospheric chlorine. Recent declines in levels of tropospheric and stratospheric chlorine that are evident in model calculations are mainly attributable to the dramatic changes in methyl chloroform described above [24, 37].

In the context of the phase-out of methyl chloroform, calculated emissions of certain other shorter-lived compounds, especially tetrachloroethene, are being combined with atmospheric measurements in order to improve understanding of atmospheric processes. The three shorter-lived chlorinated solvents considered here, trichloroethene, tetrachloroethene and methylene chloride, are used in a wide variety of industrial, commercial and consumer applications. Apart from use as chemical feedstocks, all of the applications are potentially dispersive with rapid emissions. Extensive recycling takes place both internally within the use systems themselves and externally through solvent recycling contractors. However, as for methyl chloroform, unless specific steps are taken to capture and destroy used materials, they are ultimately released into the atmosphere, so that emissions are equivalent to the amount sold. Estimates of emissions have been developed based on the audited sales which were reported in the same six geographical regions as for methyl chloroform [17, 38]. These covered production in the OECD from 1988 onwards and corresponding sales throughout the world; production outside the OECD is estimated to add about 10% to the global total. The audited sales within each region and the global emission estimates derived from these for each compound are given in Table 9. The trends in global sales over the last ten years are shown in Fig. 13 to have been relatively flat for trichloroethene and methylene chloride or to have fallen slightly in the case of tetrachloroethene. This may be contrasted with the dramatic decline of the regulated ODS, methyl chloroform.

Trichloroethene, tetrachloroethene and methylene chloride are all sufficiently reactive that the temporal or seasonal distribution of their emissions needs to be known as well as their geographical spread, if the data are to be used for modelling atmospheric chemistry and dynamics [8]. The seasonality of emissions was addressed using methylene chloride as the model compound, because – unlike the other two compounds – it is sold in many consumer applications as well as its industrial uses; therefore, any seasonal component of its use should show up most clearly. It was found that there were no discernible trends throughout the year, either overall, among the specific end uses, or in discrete geographical regions [17]. Thus, annual sales were best distributed evenly throughout the year. A recent analysis found that combustion sources were insignificant compared to industrial emissions [8]. The emissions from reported regional sales can be distributed among countries on the basis of their economic activity described by the national GDP as discussed in Sect. 2 above. Some large emitters can be located in the US Toxic Release Inventory (TRI) [39] and other databases; the rest of the national emissions have been mapped onto a 1° latitude by 1° longitude grid, using Li's population distribution within each country as described above for methyl chloroform [8]. The resultant global maps of emissions will also be posted at the RCEI web site by the end of 1998. These show, not unexpectedly, that emissions are concentrated in the industrial regions of the northern hemisphere plus the developing nations of Asia.

Table 9. Annual reported sales by region and estimated global emissions of trichloroethene, 1988–96 (in kt per year) [17, 38]

Year	Sales by region (reporting companies only)						Other 30–90°N not included in survey	Estimated Global Emissions
	90–0°S	0–30°N	30°N–90°N					
			N. America	Europe	Far East	Other 30–90°N		
1988	4	12	46	110	74	5	12	260
1989	2	10	41	104	66	3	12	235
1990	3	14	43	105	62	4	12	241
1991	3	16	34	97	47	5	10	212
1992	2	12	37	74	60	2	10	197
1993	3	15	36	90	56	3	10	213
1994	4	23	37	88	62	4	11	229
1995	5	28	40	84	69	1	11	238
1996	9	30	37	83	70	0.4	11	240

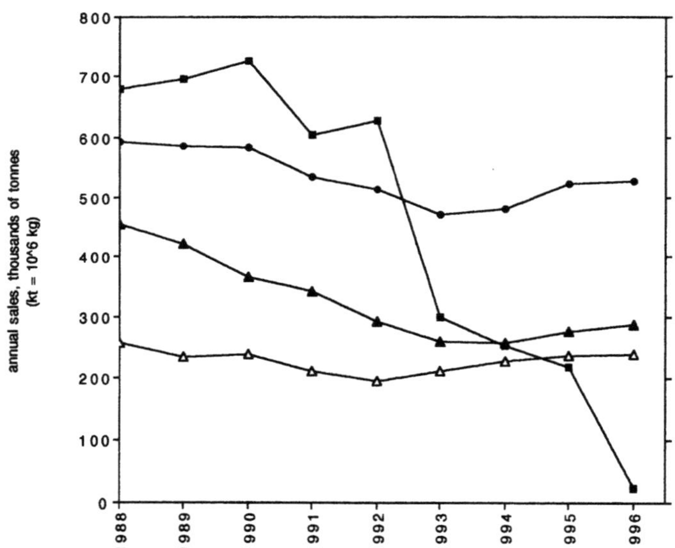

Fig. 13. Annual global sales of chlorinated solvents, 1988–1996 [5, 16, 17, 35, 38]. ■ Methyl chloroform, ● methylene chloride, ▲ tetrachloroethene, Δ trichloroethene

Table 10. Annual reported sales by region and estimated global emissions of tetrachloroethene, 1988–96 (in kt per year) [17, 38]

Year	Sales by region (reporting companies only)						Other 30–90°N not included in survey	Estimated Global Emissions
	90–0°S	0–30°N	30–90°N					
			N. America	Europe	Far East	Other		
1988	5	28	189	128	65	19	22	454
1989	3	23	173	126	55	24	20	423
1990	4	28	127	124	47	18	17	366
1991	6	23	135	116	41	7	16	342
1992	6	26	104	104	38	4	14	295
1993	9	14	86	98	39	4	12	261
1994	11	15	80	89	42	8	12	258
1995	19	20	89	82	51	4	13	278
1996	19	19	100	84	52	2	14	289

Table 11. Annual reported sales by region and estimated global emissions of methylene chloride, 1988–96 (in kt per year) [17, 38]

Year	Sales by region (reporting companies only)						Other 30–90°N not included in survey	Estimated Global Emissions
	90–0°S	0–30°N	30–90°N					
			N. America	Europe	Far East	Other		
1988	18	43	177	180	92	46	29	592
1989	16	48	161	188	95	51	28	586
1990	16	52	158	187	103	38	28	583
1991	22	51	138	173	99	27	25	534
1992	19	52	167	135	99	17	24	513
1993	14	37	98	155	123	18	22	472
1994	17	39	112	147	120	22	23	480
1995	37	52	115	149	128	16	25	523
1996	37	73	116	145	123	8	25	527

5.5 Other Halocarbons

All of the halocarbons discussed earlier in this chapter are used principally to provide effects such as refrigeration or insulation. When a chemical compound is used in this way, it is to be expected that a substantial part of the quantity manufactured will eventually be released into the environment. The emission function then accommodates the fraction released and the delay between production and release, as described in Sect. 3.2 above. In such cases, audited production and sales data are of value to provide a basis to estimate emissions rigorously.

For many materials, however, the quantity produced is not indicative of the quantity – if any – that might be emitted into the environment. Carbon tetrachloride, methyl chloride, ethylene dichloride, vinyl chloride, vinylidene chloride, tetrachloroethane and pentachloroethane are halocarbons that are not included in audited data. All have been industrially produced in substantial quantities, but they are chemical intermediates that are only emitted adventitiously during manufacturing and processing. These "fugitive" emissions vary among processes and depend, in part, on local environmental regulations.

Simmonds et al. [40] showed, on the basis of atmospheric measurements, that the quantity of carbon tetrachloride emitted during the production of CFCs 11 and 12 represented 8% of the quantity required. This probably reflects other uses for carbon tetrachloride, as it is a significantly larger loss than in the case of ethylene dichloride, for which the reported emission factors range from 0.02 to 1.2%, depending on the nature of the process [41]. Thus, the quantity of ethylene dichloride released in the USA, as reported to the TRI for 1992, was 1440 tonnes [39]. Ethylene dichloride is primarily the feedstock from which vinyl chloride and thence polyvinylchloride (PVC) is manufactured: this reference also reports that 500 tonnes of vinyl chloride have been emitted in the USA during 1992. The reported release of vinylidene chloride, precursor of polyvinylidene chloride, was 109 tonnes. Although these compounds are manufactured in quantities similar to those of the CFCs, emissions are several orders of magnitude lower, as expected for materials whose primary use is feedstock.

Only 26 tonnes of 1,1,2,2-tetrachloroethane were reported; in the past this was used to make trichloroethene, and – in a similar process – pentachloroethane was used to make tetrachloroethene. The current low emissions reflect changes in technology to more efficient and safer direct processes, in which tetrachloroethane and pentachloroethane – if they are formed at all – are not isolated.

Methyl chloride is consumed, with minimal releases, in the manufacture of silicone fluids and elastomers; an emission function of 1.4%, based on the data in TRI, was developed for it by McCulloch et al. [42]. It appears that most of the methyl chloride in the environment arises naturally [43]. The same is true of chloroform, the principal anthropogenic source of which is as a byproduct of aqueous oxidation processes, particularly paper-making and the treatment of potable water [7].

6 References

1. Prinn RG, Weiss RF, Miller BR, Huang J, Alyea FA, Cunnold DM, Fraser PJ, Hartley DE, Simmonds PG (1995) Science 269:187
2. Duafala T (1998) Uses and production of anthropogenic methyl bromide. In: Fabian P (ed) Reactive halogen compounds in the atmosphere. Handbook of Environmental Chemistry. Springer, Heidelberg, in press
3. Midgley PM, McCulloch A (1998) International regulations on halocarbons. In: Fabian P (ed) Reactive halogen compounds in the atmosphere. Handbook of Environmental Chemistry. Springer, Heidelberg, in press

4. McCulloch A, Midgley PM, Fisher DA (1994) Atmos Environ 28:2567
5. Midgley PM, McCulloch A (1995) Atmos Environ 29:1601
6. Midgley PM, McCulloch A (1997) Atmos Environ 31:809
7. Aucott ML, McCulloch A, Graedel TE, Kleiman G, Midgley PM, Li Y-F (1998) J Geophys Res, in press
8. McCulloch A, Aucott ML, Graedel TE, Kleiman G, Midgley PM, Li Y-F (1998) J Geophys Res, in press
9. McCarthy RL, Bower FA, Jesson JP (1977) Atmos Environ 11:491
10. Gamlen PH, Lane BC, Midgley PM, Steed JM (1986) Atmos Environ 20:1077
11. Midgley PM, Smythe KD (eds) (1998) Production, sales and atmospheric release of fluorocarbons through 1996. AFEAS (Alternative Fluorocarbons Environmental Acceptability Study), 1333 H St. N.W., Washington, D.C. 20005
12. Fisher DA, Midgley PM (1993) Atmos Environ 27 A: 271
13. McCulloch A (1992) Atmos Environ 26 A: 1325
14. Midgley PM, Fisher DA (1993) Atmos Environ 27 A: 2215
15. Midgley PM (1989) Atmos Environ 23:2663
16. Midgley PM, McCulloch A, Simmonds PG (1998) The phase-out of methyl chloroform and the decline in tropospheric chlorine. In: Borrell PM, Borrell P (eds) Proceedings of EUROTRAC-2 symposium '98: transport and chemical transformation in the troposphere. In press
17. McCulloch A, Midgley PM (1996) Atmos Environ 30:601
18. Ozone Secretariat to the Vienna Convention and Montreal Protocol (1997) Production and consumption of ozone depleting substances 1986–1995. UNEP (United Nations Environment Programme), Nairobi, Kenya. ISBN number: 92-807-1665-4
19. Oberthür S (1997) Production and consumption of ozone depleting substances 1986–1995, the data reporting system under the Montreal Protocol. Deutsche Gesellschaft für Technische Zusammenarbeit (GTZ), Berlin
20. UNEP (1991) Handbook for the Montreal Protocol on substances that deplete the ozone layer, 2nd edn. Ozone Secretariat to the Vienna Convention and Montreal Protocol, UNEP (United Nations Environment Programme), Nairobi, Kenya
21. Hartley DE, Prinn RG (1993) J Geophys Res 98:5183
22. Fisher DA, Duafala T, Midgley PM, Niemi C (1994) Production and emission of CFCs, halons, and related molecules. In: Kaye JA, Penkett SA, Ormond FM (eds) Report on concentrations, lifetimes, and trends of CFCs, halons, and related species. NASA Reference Publication 1339, Washington, D. C., chap 2
23. Fisher DA, Midgley PM (1994) J Geophys Res 99:16643
24. WMO (1998) Scientific assessment of ozone depletion: 1998, WMO global ozone research & monitoring project. WMO (World Meteorological Organization), Geneva, Switzerland, in press
25. Elkins JW, Thompson TM, Swanson TH, Butler JH, Hall BD, Cummings SO, Fisher DA, Raffo AG (1993) Nature 364:780
26. Cunnold DM, Fraser PJ, Weiss RF, Prinn RG, Simmonds PG, Miller BR, Alyea FN, Crawford AJ (1994) J Geophys Res 99:1107
27. Montzka SA, Butler JH, Myers RC, Thompson TM, Swanson TH, Clarke AD, Lock LT, Elkins JW (1996) Science 272:1318
28. Miller BR, Huang J, Weiss RF, Prinn RG, Fraser PJ (1998) J Geophys Res 103:13237
29. Midgley PM, Fisher DA, Kleiman G, McCulloch A (1997) Emission inventories for industrial halocarbons. In: AGU Fall Meeting Supplement, EOS Trans 78:F105
30. Fraser PJ, Oram DE, Reeves CE, Penkett SA, McCulloch A (1998) submitted to J Geophys Res
31. WMO (1985) Atmospheric ozone: 1985, WMO global ozone research & monitoring project. Report No. 16, WMO (World Meteorological Organization), Geneva, Switzerland
32. Li Y-F (1996) Global population distribution database. Canadian Global Emissions Inventory Centre, Atmospheric Environment Service, Environment Canada, Downsview, Ontario, Canada

33. McCulloch A, Midgley PM (1998) Atmos Environ 32:1571
34. Oram DE, Sturges WT, Penkett SA, McCulloch A, Fraser PJ (1998) Geophys Res Lett 25:35
35. CEFIC (1996) Worldwide sales of 1,1,1-trichlorethane (methyl chloroform) 1994–1995. CEFIC (Conseil Européen des Fédérations de l'Industrie Chimique), Ave. E. Van Nieuwenhuyse 4, B-1160 Brussels
36. Ozone Secretariat to the Vienna Convention and Montreal Protocol (1998) Report of the secretariat on information provided by the Parties in accordance with Article 7 and 9 of the Montreal Protocol. UNEP/OzL.Pro.9/ImpCom/20/3, 21 May 1998. UNEP (United Nations Environment Programme), Nairobi, Kenya
37. SORG (United Kingdom Stratospheric Ozone Review Group) (1996) Stratospheric ozone 1996. UK Department of the Environment, London
38. CEFIC (1997) Worldwide survey 1996 - methylene chloride, perchloroethylene, trichloroethylene. CEFIC, Ave. E. Van Nieuwenhuyse 4, B-1160 Brussels
39. USEPA (United States Environmental Protection Agency) (1994) EPA 1987–1992 toxic release inventory, CD-ROM version. Reference EPA 749/C-94-001. EPA, Office of Pollution Prevention and Toxics, Washington, D. C.
40. Simmonds PG, Cunnold DM, Weiss RF, Prinn RG, Fraser PJ, McCulloch A, Alyea FN, O'Doherty S (1998) J Geophys Res in press
41. USEPA (United States Environmental Protection Agency) (1978) Investigations of selected environmental pollutants: 1,2-dichloroethane. Report EPA-560/2-78-006. Washington D.C.
42. McCulloch A, Aucott ML, Benkovitz CM, Graedel TE, Kleiman G, Midgley PM, Li Y-F (1998) submitted to J Geophys Res
43. Keene WC, Khalil MAK, Erickson DJ III, McCulloch A, Graedel TE, Lobert JM, Aucott ML, Gong S-L, Harper DB, Kleiman G, Midgley PM, Moore RM, Seuzaret C, Sturges WT, Logan JA, Yevich R, Benkovitz CM, Koropalov V, Barrie LA, Li Y-F (1998) J Geophys Res, in press

CHAPTER 7

Properties, Applications and Emissions of Man-Made Methyl Bromide

T. Duafala · M. Gillis

Trical, Inc., 8770 Highway 25, Hollister, CA, USA.
E-mail: thomas@duafala.com; mattman@ix.netcom.com

Methyl bromide is both a naturally produced chemical and a man-made chemical. Methyl bromide is naturally produced by oceans, biomass burning, and plants. Anthropogenic methyl bromide is produced intentionally by man as a fumigant and unintentionally by combustion processes involving leaded gasoline and human-caused biomass burning. Approximately 70 % of man-made methyl bromide is used as a fumigant; of that, about 75 % is used is for pre-plant soil treatments, 21 % for commodity treatments, and 3 % for structural treatments. Because of international government restrictions on production and consumption, the use of methyl bromide is currently limited to 1991 levels. Because a percentage of methyl bromide is degraded during use, emissions do not equal consumption. It is estimated that approximately 63,418 metric tons of methyl bromide are emitted per year from all anthropogenic sources.

Keywords: Methyl bromide, Bromomethane, Fumigant, Emissions.

1 Introduction

Methyl bromide, or bromomethane, is a volatile organic chemical that has both anthropogenic and natural sources (see Singh and Fabian, this Vol.). Anthropogenic sources include both intentional and unintentional production from manufacturing processes or combustion, such as forest fires [1]. Natural production takes place in oceans and by plants [2, 3]. Methyl bromide has been in

The Handbook of Environmental Chemistry Vol. 4 Part E
Reactive Halogen Compounds in the Atmosphere
(ed. by P. Fabian and O. N. Singh)

anthropogenic use for a variety of purposes for over 50 years. Methyl bromide was used in the 1940s as a fumigant to treat foodstuffs and stored products. During the same period, it was also used as a fire extinguishing agent, a refrigerant, and an insecticide for exterminating lice and bedbugs [4]. These latter uses were discontinued due to methyl bromide's high acute human toxicity. The primary use today is as a fumigant for controlling pests in the stored product, structural, and agricultural industries. Regardless of the use, containment is required for effective control. The degree of containment and the treated matrix determine the percentage of applied methyl bromide lost to the atmosphere.

2 The Important Properties of Methyl Bromide

Methyl bromide has certain physical properties that make it a good fumigant. Table 1 lists some of its physical properties [5]. At ambient temperatures, methyl bromide is a colorless and odorless gas. The boiling point at 3.6 °C (37 °F) allows

Table 1. Methyl bromide physical properties

Boiling Point	3.6 °C
Freezing Point	-93 °C
Molecular weight	94.94
Specific gravity: gas (air = 1)	3.27 at 0 °C
Specific gravity: liquid (water at 4 °C = 1)	1.732 at 0 °C
Latent heat of vaporization	61.52 cal/g
Flammability limits in air	Non-flammable
Solubility in water	1.34 g/100 ml at 25 °C
Pertinent chemical properties	Powerful solvent of organic materials. especially natural rubber. When pure, non-corrosive to metals. Reacts with aluminum, magnesium, zinc, alkali metals and strong bases
Method of evolution as fumigant	From steel cylinders under natural or added pressure. Also dispensed from 1 lb cans
Commercial purity	~99.5%
Vapor pressure at 0 °C	690 mm Hg
Vapor pressure at 10 °C	1006 mm Hg
Vapor pressure at 20 °C	1390 mm Hg
Vapor pressure at 25 °C	1610 mm Hg
Weights and volumes of liquid at 0 °C	1 lb. (avdp) at 0 °C has volume of 261.9 ml 1 US gallon weighs 14.44 lb (6.550 kg) 1 Imperial gallon weighs 17.32 lb (7.856 kg) 1 kg has volume of 577.36 ml 1 l weighs 1.732 kg
Henry's law constant at 20 °C	0.533 kPa m^3/mol

containerized methyl bromide to be handled as a liquid under relatively low pressures, but to change phase to a gas after it is applied. Relative to other pesticides, it has a high vapor pressure: 1390 mm Hg at 20 °C. The vapor pressure provides a driving force to move the fumigant through the treated matrix rapidly and extensively enough to provide uniform pest control. Methyl bromide liquid is contained in cylinders or cans, usually under its own pressure of about 2 atm.

3 Applications and Usage of Methyl Bromide

For the past 30 years the primary anthropogenic use of methyl bromide has been as a fumigant. Methyl bromide is a broad-spectrum biocide; it is easy to handle and has excellent penetration properties. These properties have helped to maintain the popularity of methyl bromide, even though it is acutely toxic to humans. Methyl bromide has also allowed for continuous cropping of agricultural soils without loss of productivity. Because of its properties and reasonable cost, methyl bromide is used as a fumigant in both developed and developing countries.

Approximately 75% of intentionally produced methyl bromide is used as a pre-plant soil fumigant for the control of soil-borne diseases, nematodes, weeds and weed seeds, and plant-destroying insects [6]. Soil fumigation with methyl bromide was developed in the early 1960s, and by1965 nearly 100% of all strawberries in California were treated with this pesticide [7]. The development of easy-to-use application equipment was largely responsible for the rapid growth in use of methyl bromide. When used as a pre-plant soil fumigant, methyl bromide is either injected into the soil at a depth of 6–10 in under polyethylene tarp or introduced between the soil surface and a polyethylene tarp. In the United States, deep injection at a depth of 18–30 in. without tarp is also permitted. The method of application depends on the crop to be grown and the part of the world where the application is taking place. For example, injection into soil under tarp is the common method of application in the United States, whereas application between the soil surface and a polyethylene tarp is most common in Europe. With the latter method of application, methyl bromide can be introduced either in a heated vapor state or as a liquid. Emissions from soil can be controlled either by increasing the depth of application or by covering soils with polyethylene film. In most areas of the world, polyethylene tarps are the most common and effective method of containment, particularly where soil injection is not possible.

Approximately 24% of intentionally produced methyl bromide is used in commodity, quarantine, and space treatments for the control of insects, rodents and other harmful pests [6, 8]. Treatments of this type were developed in the 1940s [9]. One very important use is the treatment of dried fruits, fresh fruits, nuts, and grains, to preserve the storage quality of the product or to prevent the importation of unwanted pests into the importing country. These treatments take place in fumigation chambers, warehouses, and vacant structures, and under gas impermeable tarps. Because of the ease of application and containment, methyl bromide is the treatment of choice in many developing countries.

Although these types of treatment account for only 21% of methyl bromide use, they are of critical importance, since they result in the preservation of valuable food supplies, and are often required by importing countries as part of their quarantine program. Structural (space) fumigation accounts for 3% of its use [6].

A small amount of methyl bromide is also used as a chemical intermediate in several manufacturing processes. The amount used has remained relatively constant over the period 1984–1992 and has accounted – on average – for 5.9% of total production [8]. This number is not expected to have changed in the last 6 years.

4
Regulatory Activity

Because of methyl bromide's use as a pesticide and its acute toxicity, it is highly regulated in all countries that permit its use. For example, in many countries it can only be applied by trained applicators. The Netherlands began a 10-year phase-out of methyl bromide in 1981; use of methyl bromide has not been allowed since 1991 [10]. In 1991, methyl bromide came under scrutiny as an ozone-depleting chemical and, for that reason, several countries have also restricted its use or scheduled it for phaseout. In addition to actions taken by individual countries, methyl bromide is now regulated by the Montreal Protocol, an international treaty for the protection of the ozone layer [11] (see Midgley and McCulloch: International Regulations on Halocarbons, this Vol.). Under the Montreal Protocol, production of methyl bromide is frozen at 1991 levels; it is scheduled for phase-out by 1 January 2005 in developed nations and by 2015 in developing nations. Developed nations will be required to implement interim reductions in production and import of 25% in 1999, 50% in 2001, and 70% in 2003. Developing nations are required to implement a reduction in production and import of 20% in the year 2005 [12]. Under the U.S. Clean Air Act of 1990, methyl bromide was scheduled for phaseout in the United States by 1 January 2001 [13]. In an effort to create equity between the U.S. and other developed nations, the U.S. Congress amended the Clean Air Act in the Fall of 1998 to bring the U.S. in line with the Montreal Protocol phase-out schedule.

5
Production of Methyl Bromide

Methyl bromide production is primarily limited to areas of the world which have a ready supply of bromine. For example, in the United States, methyl bromide is produced in the State of Arkansas with bromine that comes from deep wells. In Israel, methyl bromide is produced with bromine from the Dead Sea. It is either produced directly from the reaction of methanol with hydrobromic acid or as a co-product of brominated flame-retardant production. Methyl bromide is produced primarily in the United States, Israel, France and Japan, see Table 2 for a list of manufacturers.

Table 2. Listing of methyl bromide-reporting companies, inclusive of related subsidiaries and/or joint ventures that have reported production and release data

1. Albemarle Corporation (formerly Ethyl Corporation) (US)
   Ethyl S.A. (Belgium)
2. Association of Methyl Bromide Industry Japan (Japan)
   Sanko Kagaku Kogyo Co., Ltd.
   Teijin Chemicals, Ltd.
   Nippoh Chemicals Co., Ltd.
   Dohkal Chemical Industry Co., Ltd.
   Nippon Kayaku Co., Ltd.
   Ichikawa Gohsei Chemical Co., Ltd.
3. Atochem S.A. (France)
4. Dead Sea Bromine Compounds, Ltd. (Israel)
5. Great Lakes Chemical Company (US)
   Great Lakes Chemical (Europe)
   Great Lakes Chemical (UK)

Table 3. Total production and sales of methyl bromide on an annual basis for the period 1984 through 1992 from reporting companies (metric tons)

Year	Production	Sales
1984	44,856	45,572
1985	48,013	48,273
1986	48,754	50,455
1987	56,224	55,690
1988	61,546	60,610
1989	64,628	62,570
1990	61,724	66,644
1991	70,022	71,260
1992	69,882	71,585
Totals	525,649	532,659

Production data generated by the Methyl Bromide Global Coalition covers the period 1984–1992 [8]. Industry records prior to 1984 were not sufficiently complete to allow for the development of earlier production data. The major use of methyl bromide as an agricultural pesticide did not start to develop until the 1960s; therefore it can be presumed that production before this time was limited. Industry data show that growth in production occurred at 8% per year between 1984–1992 see (Table 3).

The methyl bromide industry did not collect production data after 1992, since methyl bromide's production is frozen in the United States under the US Clean Air Act and in other developed nations under the Montreal Protocol, with production limited to 1991 levels in both cases. Parties to the Montreal Protocol are now required to report both production and consumption data. The first report to the Parties was issued by the United Nations Environment Program (UNEP)

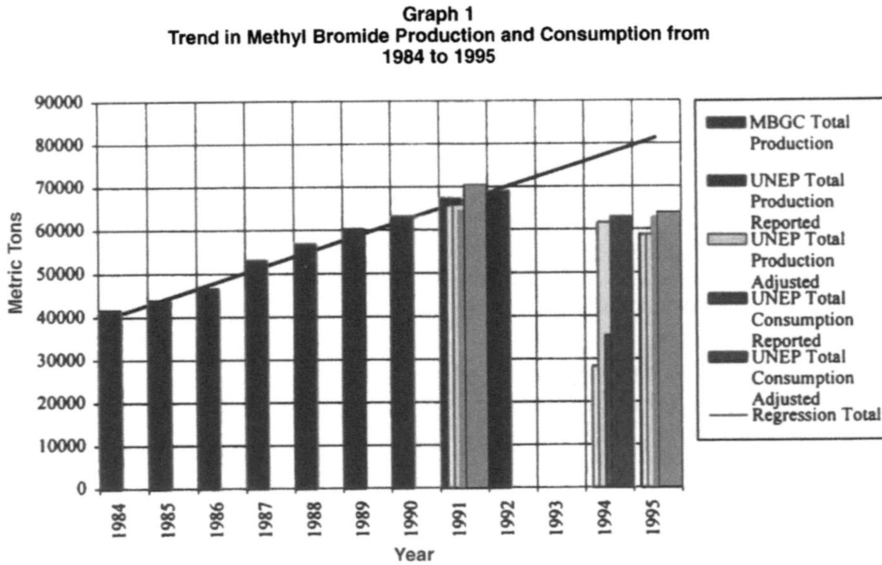

Fig. 1. Trend in methyl bromide production and consumption from 1984 to 1995

in 1997 (see Tables 4 and 5; [14]). When consumption data are adjusted for nations such as Japan and Israel, who did not report for 1994, it can be seen that consumption of methyl bromide decreased slightly from 1991–1995 (see Fig. 1). This would be consistent with international controls on production. Unfortunately, production data as reported to UNEP are not consistent with consumption data for the years 1994 and 1995 and suggest reduction in use or depletion of stockpiled product. It is unlikely that there was much stockpiled methyl bromide, since industry would have had only 1 year, 1993, to produce excess product for use in subsequent years. The most probable reason for this discrepancy is incomplete reporting by producing countries. The methyl bromide industry reported that in 1992, 69,882 metric tons of methyl bromide were produced and 71,585 metric tons were consumed (see Table 3). UNEP reported for 1995 that 58,727 metric tons of methyl bromide were produced and 62,503 metric tons were consumed (see Tables 4 and 5).

Methyl bromide is also produced as an inadvertent by-product of other manufacturing processes. Amoco reported that it released approximately 357 metric tons of methyl bromide in 1995 from its US facilities [15]. Amoco is just one of several US companies that produces methyl bromide inadvertently. In cooperation with US EPA, this group of companies has agreed to reduce emissions of methyl bromide from 1991 levels by 20% by the end of 1997 and by 85% by the end of the year 2000. According to the Toxic Release Inventory of the United States, industrial emissions, which also include some fumigation releases, were 1,443 metric tons in 1991 and 1,163 metric tons in 1995. No good information is available on total emissions from inadvertent production outside of the United States.

Table 4. Data on production of methyl bromide (Annex E, Group I) (Data was converted from ODP tons to metric tons)

Year	1991	1994	1995
Non-Article 5 Parties			
Asia			
Israel	24,508		24,213
Japan	5627		5270
Eastern Europe			
Ukraine	0	0	1402
Western Europe and others			
France	3648		3580
USA	31,827	28,217	23,942
Article 5 Parties			
Asia			
China			285
India	123		
Eastern Europe			
Romania		22	35
Totals	65,733	28,238	58,727

Table 5. Data on consumption of methyl bromide (Annex E, GroupI) (Data was converted from ODP tons to metric tons)

Year	1991	1994	1995
Non-Article 5 Parties			
Africa			
South Africa	1265	1002	1007
Asia			
Japan	6107		5972
Israel	20,512		20,443
Eastern Europe			
Belarus	- 40	0	25
Bulgaria	0	15	12
Czech Republic	10	8	18
Hungary	53	73	53
Latvia	25		20
Lithuania	55	37	52
Poland	200		
Russian Federation	0	1738	2383
Ukraine	0	0	650
Western Europe and others			
Australia	703	0	493
Austria	3	2	
Canada	200	215	125
Finland	8	8	

Table 5 (continued)

Year	1991	1994	1995
Western Europe and others			
New Zealand	150	148	128
Norway	10	7	8
Sweden	27	N.R.	
Switzerland	0	27	23
USA	29,175	21,080	20,807
Article 5 Parties			
Africa			
Algeria	2	2	2
Cameroon	0	23	25
Cote D'Ivoire	0		15
Egypt	80	150	450
Ethiopia	22		22
Kenya	330	360	225
Malawi	113	203	163
Morocco	222	883	1298
Mozambique	7	8	10
Senegal	270	350	332
Sudan	10	-10	-17
Tanzania, United Rep. of 7	0	0	
Tunisia	N.R.	10	20
Zambia	7	40	53
Zimbabwe	628	628	677
Asia			
China			658
India	188		
Indonesia	135		50
Iran, Islamic Rep. of	60	122	25
Jordan	112	333	300
Korea, Rep. of	1412	0	0
Kuwait	N.R.	20	0
Malaysia	0	38	57
Myanmar		0	17
Papua New Guinea	0		2
Philippines	35	35	17
Sri Lanka	0		7
Syrian Arab Republic	5		
Thailand	N.R.		567
Turkey	493		702
Vietnam	460	140	550
Eastern Europe			
Croatia	32	28	28
Georgia	158		
Romania		20	33
Slovenia	0	0	2
The Former Yugoslav Rep. of Macedonia	20		20

Table 5 (continued)

Year	1991	1994	1995
Latin America and the Caribbean			
Argentina	N.R.	282	N.R.
Brazil	N.R.	940	
Chile	182	198	295
Colombia	218	72	
Costa Rica	460	500	550
Cuba	110	45	
Dominica	5		
Dominican Republic	57	82	115
Ecuador	N.R.	N.R.	65
El Salvador	3	5	25
Guatamala	0	95	418
Guyana			5
Mexico	N.R.	5422	2397
Panama	8	2	0
Peru	2	2	2
Uruguay	5	23	25
Venezuela		3	17
Western Europe and others			
Malta	40	112	62
Totals	64,360	35,527	62,503

Other anthropogenic sources include biomass burning and use of leaded gasoline. It is estimated that anthropogenic biomass burning contributes 20,000 metric tons of methyl bromide emission per year [16], and leaded gasoline contributes less than 5,000 metric tons per year [17].

6 Emissions from Anthropogenic Uses

A discussion on methyl bromide emission is complex, because of the multiple anthropogenic sources and partial destruction of the chemical in many of its anthropogenic uses. It is assumed that all of the methyl bromide produced from biomass burning and leaded gasoline combustion is emitted into the atmosphere. The only issue to be resolved with respect to emissions from biomass burning and leaded gasoline is the amount of methyl bromide produced. In recent years estimates of the amount of methyl bromide produced from combustion and emitted to the atmosphere have decreased [15, 16].

Inadvertent emissions from industrial production are even more difficult to quantify. Some countries, such as the United States, have strict reporting requirements on these emissions, but similar information is not available from many countries. Based on US reporting figures from the Toxic Release Inventory, emissions from inadvertent production should not be a significant portion of the total production of methyl bromide.

Emissions of methyl bromide from use as a fumigant are also difficult to quantify. Emissions vary depending on the type of application method used and the matrix being treated. In soil treatments it is estimated that emissions range from 21 to 89% of applied chemical [17–21]. The current best estimate of overall average loss from all types of soil treatment is 50 ± 10% of the fumigant applied [17].

Regardless of how methyl bromide is applied to the soil, emissions are generally greatest during the first 24 h following application. As the methyl bromide disperses in the soil, methyl bromide near the surface is lost to the atmosphere and the rest becomes more diffuse, causing the concentration of the gas in the soil to decrease. Since emissions are largely a function of methyl bromide concentration at the soil surface, emission rates are generally reduced as elapsed time from application increases. Changes in soil temperature, moisture, and atmospheric pressure will cause fluctuations in emission rate [18, 21].

Cultural practices prior to fumigation can affect the soil's physical and chemical properties that affect fumigant movement. Growers preparing their soil for fumigation will often pre-irrigate and work the ground extensively, in order to bring it into a condition for optimal fumigant dispersion. Organic material from previous crop residues can absorb methyl bromide and affect distribution.

As mentioned previously, tarps provide the simplest and most effective way to reduce methyl bromide emissions from fumigated fields. Tarps are typically made of low-density polyethylene or a combination of low-density/high-density polyethylene. All of these films provide effective barriers for retaining methyl bromide for purposes of efficacy, but – depending on the film used – will result in different emission factors. With the need to reduce emissions because of concerns over ozone and offsite exposure, films with greater barrier properties are being developed. Some virtually impermeable films result in negligible permeation of methyl bromide through the film. Based upon laboratory studies, some of the new films may reduce permeation compared with polyethylene films by 10 to 100 times, although emissions under field conditions are reduced by a much smaller factor. Until now, practical and economic concerns have limited implementation of these films commercially, but ongoing research may ultimately bring them into commercial practice.

Structural and commodity fumigations have fewer variables that impact emissions than soil fumigation. It is assumed that nearly 100% of the fumigant applied in structural fumigation is emitted into the atmosphere. The purpose of structural fumigation is to treat a structure rather than its contents. Generally these structures are empty or nearly empty. Therefore, there are limited sorptive sites for removing the methyl bromide. Thus, when the fumigation is complete and the structure is vented, a significant portion of the methyl bromide applied is emitted to the atmosphere.

It is also estimated that most of the methyl bromide applied in commodity fumigation is emitted into the atmosphere. Commodities are generally fumigated within an enclosed space such as a fumigation chamber or the cargo hold of a ship. After the required exposure period, the enclosure is vented, releasing the methyl bromide into the atmosphere. Some of the methyl bromide is absorbed into the commodity, with the amount absorbed varying according to the com-

Table 6. Summary of anthropogenic emissions of methyl bromide in metric tons per year (based on adjusted 1995 UNEP consumption data)

Source	Percent ermitted	Best estimate of emission (metric tons)	Range
Soil fumigation	50 ± 10	23,833	19,066 to 28,599
Space fumigation	100	1907	1907
Commodity fumigation[a]	90–99	12,679	12,012 to 13,213
Gasoline	100	5000	0 to 10,000
Biomass burning	100	20,000	10,000 to 40,000
Chemical intermediate	0	0	
Inadvertent industrial production[b]	?	?	
Total emissions		63,418	42,984 to 93,718

[a] Amounts not emitted are estimated from inorganic bromide residue analysis [24].
[b] No values are available for worldwide inadvertent production.

modity. Some commodities absorb larger amounts of methyl bromide than others do. The methyl bromide that is absorbed either slowly desorbs or methylates compounds in the commodity, resulting in degradation of the methyl bromide. Although fumigation of some commodities high in oils will result in absorption and degradation of more than 10% of the methyl bromide applied, for other commodities, most of the methyl bromide absorbed is desorbed, resulting in greater than 95% emissions. For example, in one study it was calculated that walnuts retained 8–20% methyl bromide as inorganic residue, and stone fruits retained 2–3% methyl bromide as inorganic residue [24].

In view of the aforementioned difficulties related to the complex nature of methyl bromide's sources, uses, and degradation processes during use, the best estimate of methyl bromide's emissions from all anthropogenic activities is 63,418 metric tons/year based on 1995 figures (see Table 6). Uncertainties in the best estimate of the value result in a range of 42,984 to 93,718 metric tons/year.

7 References

1. Ennis CA (ed) (1994) World Meteorologic Organization Global Ozone Research and Monitoring Project,Report No 37, Scientific assessment of ozone depletion
2. Anbar AD, Yung YL, Chavez FP (1996) Global Biogeochem Cycles 10:175
3. Gan J, Yates SR, Ohr HD, Sims JJ (1998) Geophys Res Lett (in press)
4. Butler ECB, Perry KMA, Williams JRF (1945) Br J Ind Med 2:30
5. Monro HAU (1969) Manual of fumigation for insect control, 2nd edn. Food and Agriculture Organization of the United Nations, Rome
6. UNEP Methyl Bromide Technical Options Committee (1994) Montreal Protocol on Substances that Deplete the Ozone Layer
7. Wilhelm S, Paulus AO (1980) Plant Dis 64:265
8. Thornton G (ed) (1994) Methyl bromide annual production and sales for the years 1984–1992. Accountants' compilation report

9. Brown WB (1954) Pest Infestation Research. Her Majesty's Stationery Office, London
10. De Barro PT, Edwards B (1995) Strawberry production in the Netherlands without methyl bromide. In: Banks HJ (ed) Agricultural production without methyl bromide – four case studies. CSIRO Division of Entomology, Australia, p 29
11. UNEP (1996) Handbook for the international treaties for the protection of the ozone layer, 4th edn. Secretariat for The Vienna Convention for the Protection of the Ozone Layer and the Montreal Protocol on Substances that Deplete the Ozone Layer, Nairobi, Kenya
12. UNEP (1997) Report of the ninth meeting of the parties to the Montreal Protocol on substances that deplete the ozone layer
13. United States Environmental Protection Agency (1993) Fed Regist 58:65018
14. UNEP (1997) Production and consumption of ozone-depleting substances, 1986–1995. Report to the Secretariat by the Parties to the Montreal Protocol
15. Amoco Corporation (1996) Environmental releases. Internet site http://www.amoco.com
16. Andreae MO, Atlas E, Harris CW, Helas G, de Kock A, Koppmann R, Maenhaut W, Manö S, Pollock WH, Rudolph J, Scharffe D, Schebeske G, Welling M (1996) J Geophys Res 101: 23603
17. Methyl Bromide Global Coalition (1997) Methyl Bromide State of the Science Workshop Summary, Monterey, California
18. Yates SR, Ernst FF, Gan J, Gao F, Yates MV (1996) J Environ Qual 25:192
19. Yates SR, Gan J, Ernst FF, Mutziger A, Yates MV (1996) J Environ Qual 25:184
20. Yates SR, Gan J, Ernst FF, Wang D (1996) J Environ Qual 25:892
21. Yates SR, Gan J, Wang D, Ernst FF (1997) Environ Sci Technol 31:1136
22. Yagi K, Williams J, Wang N-Y, Cicerone RJ (1995) Science 267:1979
23. Seiber JN, Knuteson JA, Woodrow JE, Wolfe NL, Yates MV, Yates SR (eds) (1997) Fumigants environmental fate, exposure, and analysis, 1st edn. American Chemical Society, Washington, DC
24. Leesch J (1998) USDA-ARS Horticultural Crops Research Lab (personal communication)

CHAPTER 8

International Regulations on Halocarbons

P. M. Midgley[1] · A. McCulloch[2]

[1] M & D Consulting, Ludwigstraße 49, D-70771 Leinfelden, Germany.
E-mail: paulinemidgley@compuserve.com
[2] ICI Chemicals & Polymers Ltd., Safety & Environment Department, PO Box 13, The Heath, Runcorn, Cheshire, WA7 4QF, UK. *E-mail: archie_mcculloch@ici.com*

The Montreal Protocol on Substances that Deplete the Ozone Layer was a landmark in international environmental regulation. This chapter describes the development of understanding of atmospheric science that led to the initial Protocol, agreed in 1987, and the subsequent developments that shaped its revisions. Eleven years later, both industrial statistics and, perhaps more significantly, atmospheric measurements clearly show compliance with the Protocol, and indicate that it is beginning to achieve its enviromental aim. The framework for regulations that now cover: chlorofluorocarbons (CFCs), halons, carbon tetrachloride, methyl chloroform (1,1,1-trichloroethane), methyl bromide, hydrochlorofluorocarbons (HCFCs) and hydrobromofluorocarbons (HBFCs) is described. The regulations themselves, which treat developing and developed countries differently, are examined together with limits to the scope for further changes that may be required to mitigate future environmental impact. Finally, the differences between the Montreal Protocol and the recently developed Kyoto Protocol on greenhouse gases are examined. The latter seeks to control emissions of greenhouse gases, such as carbon dioxide, hydrofluorocarbons (HFCs) and perfluorocarbons (PFCs).

Environmental regulations that are not specifically directed at halocarbons are not addressed in this chapter.

Keywords: Climate change, Kyoto Protocol, Montreal Protocol, Ozone depletion.

The Handbook of Environmental Chemistry Vol. 4 Part E
Reactive Halogen Compounds in the Atmosphere
(ed. by P. Fabian and O. N. Singh)

Abbreviations

AFEAS	Alternative Fluorocarbons Environmental Acceptability Study
AGWP	Absolute Global Warming Potential ($Wm^{-2}yrppb^{-1}$)
CCOL	Coordinating Committee on the Ozone Layer
CFC	Chlorofluorocarbon
CMA	Chemical Manufacturers Association
DoE	Department of the Environment (UK)
EESC	Equivalent Effective Stratospheric Chlorine (concentration)
GWP	Global Warming Potential, relative to a constant value for CO_2
HCFC	Hydrochlorofluorocarbon
HFC	Hydrofluorocarbon
IPCC	Intergovernmental Panel on Climate Change
ITH	Integration Time Horizon
NAS	National Academy of Sciences (USA)
NASA	National Aeronautical and Space Administration (USA)
NRC	National Research Council (USA)
ODP	Ozone Depletion Potential
ODS	Ozone Depleting Substance
OECD	Organisation for Economic Co-operation and Development
PFC	Perfluorocarbon
SORG	Stratospheric Ozone Review Group (UK)
TEAP	Technical and Economic Assessment Panel (for the Parties to the Montreal Protocol)
UN	United Nations
UNEP	United Nations Environment Programme
UNFCCC	United Nations Framework Convention on Climate Change
WMO	World Meteorological Organisation

1 Introduction and Background

CFCs 11 and 12 (trichlorofluoromethane, CCl_3F, and dichlorodifluoromethane, CCl_2F_2) were introduced in the 1930s as replacements for toxic and flammable refrigerants. Production and emissions remained comparatively low through to the 1950s but then increased rapidly (see Fig. 1 in the chapter on Production, Sales and Emissions of Industrial Halocarbons in this volume) with the spread of refrigeration in the developed world and as new uses, such as aerosol spray cans, were developed. By the early 1970s, CFCs 11 and 12 had become ubiquitous trace constituents of the troposphere [1]. This prompted the manufacturers of CFCs to start a programme to investigate the possible effects of CFCs on the environment and the Fluorocarbon Panel was set up in 1972 under the auspices of the Manufacturing Chemists Association (now called the Chemical Manufacturers Association, CMA) in Washington. The original aim was to assess the smog-forming potential of CFCs, which has since been found to be insignificant. However, the course of the programme was altered in 1974 when, based on Lovelock's work [1, 2], Molina and Rowland propounded their hypothesis of ozone depletion by CFCs [3].

The essence of the hypothesis was that, because of their exceptionally high chemical stability, CFCs emitted at the earth's surface would be totally stable in the troposphere and would diffuse unchanged to the stratosphere, where they would photolyse under the action of the sun's UV radiation to produce Cl atoms. These, in turn, would react with ozone to form chlorine monoxide radical ($ClO^{\bullet}$) and oxygen. In the original hypothesis, $ClO^{\bullet}$ was thought to combine with O (odd oxygen, which is in dynamic equilibrium with ozone and molecular oxygen in the stratosphere) and regenerate Cl atoms. Each cycle would generate two oxygen molecules at the expense of one ozone molecule and an odd oxygen atom. In effect, chlorine atoms resulting from the photolysis of CFCs would increase the destruction of ozone that was already taking place by Cl atoms arising from naturally occurring chlorocarbons in the stratosphere. Owing to the cyclic nature of the reaction, each Cl atom could destroy many ozone molecules before it reacted with other species to form a stable and inactive molecule, such as HCl.

That was the basic hypothesis but, at that time, no ozone depletion had been observed and mathematical models of the atmosphere were incapable of describing all of the processes consistently [4]. Throughout the 1970s and early 1980s, the scientific community strove both to detect trends in stratospheric ozone [5, 6], and to improve the models of atmospheric processes [7, 8]. Numerous assessments were conducted, but no firm conclusions were reached [9–12].

In the meantime the releases of CFCs 11 and 12 continued to grow [13, 14], as did releases of other compounds that could be transported to the stratosphere and decompose there to release chlorine or bromine: CFC-113 (1,1,2-trichlorotrifluoroethane, CCl_2FCClF_2), CFC-114 (1,1,2,2-tetrafluorodichloroethane, $CClF_2$ $CClF_2$), CFC-115 (chloropentafluoroethane, CF_3CClF_2) [15], Halon-1211 (bromochlorodifluoromethane, $CBrClF_2$), Halon-1301 (bromo trifluoromethane, $CBrF_3$) [16], carbon tetrachloride (CCl_4) and methyl chloroform (1,1,1-trichloroethane, CCl_3CH_3) [17, 18] all showed growth, although for many compounds this was not documented rigorously until comparatively recently, if at all.

The growth in emissions was reflected in growth in atmospheric concentrations of the principal compounds that was sufficiently alarming to set regulations in process, notwithstanding the inability of atmospheric models to agree or real ozone depletion to be detected [19].

In the mid 1970s, the widespread use of CFCs in aerosols was banned in USA; this resulted in an immediate reduction in emissions, but the long term trend of releases remained positive. In contrast, production was capped at the then current capacity in Europe, with a requirement to reduce the quantities used in aerosol propulsion by 30% [20, 21]. This form of regulation – controlling total production and consumption, rather than each end use – was subsequently adopted in the Montreal Protocol and its revisions.

In 1981 there was still no evidence that the ozone layer was being affected, but – with the expectation that it could become depleted – the United Nations Environment Programme set up the "Ad Hoc Working Group of Legal and Technical Experts for the Elaboration of a Global Framework Convention for the

Protection of the Ozone Layer", with the aim of securing a general treaty to tackle ozone depletion. This was finally agreed upon in Vienna in 1985 as the Convention for the Protection of the Ozone Layer.

Nations agreed to take "appropriate measures ... to protect human health and the environment against adverse effects resulting or likely to result from human activities which modify or are likely to modify the Ozone Layer;" but the measures were unspecified. The main thrust of the Convention was to encourage research, cooperation among countries and exchange of information. Twenty-eight nations signed the Convention and it has been subsequently ratified by 166 [22]; it provided the framework for the Montreal Protocol and specified procedures for its amendment.

The Vienna Convention set an important precedent; for the first time nations agreed in principle to tackle a global environmental problem before its effects were felt – or even scientifically proven. One factor that helped here was the fact that there are relatively few producers of ozone-depleting substances, even though the substances have a wide range of uses across society (see chapter on Properties and Applications of Industrial Halocarbons in this volume, Sect. 1). This meant that those drafting the treaty could envisage controls on particular substances, rather than controls on society's activities. In this respect, ozone-depleting substances are very different from greenhouse gases; the principal greenhouse gases, carbon dioxide (CO_2) and methane (CH_4), are released as byproducts of societal activities, such as energy conversion and agriculture. Consequently, the climate change controls are directed at emission, rather than production and consumption (see Sect. 3).

2 The Montreal Protocol on Substances that Deplete the Ozone Layer

At the same time as the legal and technical experts were developing treaties, the scientific experts in the Coordinating Committee on the Ozone Layer (set up in 1977) were reviewing results of atmospheric measurements and the models using them, and developing projects to extend understanding of ozone layer behaviour. The first real evidence for ozone depletion came from Farman et al. [23] who, in 1985, linked severe seasonal ozone depletion in the Antarctic to the growth in chlorine from CFCs in the Antarctic stratosphere. This paper and the work that followed (reported in [24] and [25]) was instrumental in promoting the agreement that was reached on specific measures to be taken. The Montreal Protocol on Substances that Deplete the Ozone Layer was signed by twenty four countries in September 1987; it has subsequently been ratified by 165 countries.

The Protocol, which came into force on 1 January 1989, is a flexible instrument; the provisions must be modified in the light of a virtually continuous scientific review process that reported to the Parties in 1989, 1991 and 1994 [26–28]]. Reviews of the technologies available for providing substitutes for ODS occur with similar frequency [29, 35], together with reviews of the possible effects of ozone depletion [30]. The Protocol also contains clauses to cover the special circumstances of several groups of countries, especially developing countries with low consumption rates that do not want the Protocol to hinder their development. As a result, regulations have evolved since 1989 as the scientifically driven requirements have changed and as the political and societal needs of countries have changed.

For the developed world, Article 2 of the original version of the Protocol set out to control national production and consumption of specific CFCs (11, 12, 113, 114 and 115) and halons (1211, 1301 and 2402) as two distinct groups: the CFCs were to be reduced by the year 1998 to 50% of their level in 1986, and production and consumption of halons were to be frozen at their 1986 levels in 1993. In both cases the different potency for ozone depletion of substances within each group was taken into account, using the "Ozone Depletion Potential" (ODP) of each substance as a multiplier of the masses produced or consumed. These ODPs were based on model analyses performed for the 1989 Assessment [26]; current values are shown in Table 1 in Properties and Applications of Industrial Halocarbons (this volume). Countries in the developing world, that have per capita consumption less than 0.3 kg $year^{-1}$, operate under Article 5 of the Protocol which allows for a ten year delay in meeting the control measures of Article 2.

2.1 Amendments and Adjustments to the Protocol

2.1.1 *London 1990*

The CFCs controlled in the original version of the Protocol have long atmospheric lifetimes, of the order of a few decades to several centuries. Consequently, their atmospheric concentrations will be maintained by comparatively modest emissions. In their second report, the UK's Stratospheric Ozone Review Group calculated that a 77% reduction in emissions of CFC-11 and an 85% reduction in the emissions of CFC-12 would be required, simply to stabilise atmospheric concentrations at 1989 levels [31]. Furthermore, the increases in concentration arising from production that were still allowed under the provisions of the Montreal Protocol current at the end of the 1980s were not trivial; CFC-12 levels, for example, could have doubled by 2050 had the Protocol not been changed.

At the same time, it became generally apparent that other compounds were capable of being transported into the ozone layer and augmenting ozone depletion by releasing chlorine there. Carbon tetrachloride (tetrachloromethane, CCl_4) is used principally as a raw material for the production of CFCs 11 and 12 and, during the course of its manufacture and use, some of it leaks into the atmosphere. It can also be emitted as a byproduct of procedures that involve chlorination of organic material with molecular chlorine. It is not a significant product of drinking water chlorination. The long atmospheric lifetime of carbon tetrachloride (42 years [28]) and the fact that 92% of its mass is chlorine made it an important ozone depleting substance, even though the quantities released were smaller than CFC releases.

Methyl chloroform (1,1,1-trichloroethane, CCl_3CH_3) has a much shorter atmospheric lifetime than the CFCs; the best estimate in 1989 was 6.3 years [26] and the current best estimate is 4.8 years [32]. Nevertheless, because of the large quantities emitted, its tropospheric concentration was greater than that of carbon tetrachloride [26]. A significant part of the tropospheric concentration (over 10%) could be expected to reach the stratosphere and deplete ozone there.

There were also releases of hydrochlorofluorocarbons (HCFCs) to consider. One of these, HCFC-22 (chlorodifluoromethane, $CHClF_2$), had been used as a refrigerant for many years [33] and in 1987 had an atmospheric concentration of 100 ppt (parts per trillion, 100 in 10^{12}) [26]. The concern, at the end of the 1980s and subsequently, was that removing the option to use CFCs would result in a rapid and sustained increase in the use of HCFC.

One conclusion of the 1989 Scientific Assessment of Stratospheric Ozone [26] was as follows: "To return the Antarctic ozone layer to levels approaching its natural state, and hence to avoid the possible ozone dilution effect that the Antarctic ozone hole could have at other latitudes, one of a limited number of approaches is a complete phase-out of all fully halogenated CFCs, halons, carbon tetrachloride, and methyl chloroform, as well as careful consideration of the HCFC substitutes". The theme was echoed in the third report of the SORG, in which timing of the CFC phase-out (a total cessation of emissions by 2000) was suggested, and a similar warning was sounded about HCFCs: "Substitution of CFCs by HCFCs in other than modest proportion, depending on the lifetime of the particular HCFC used, could both increase the peak chlorine loading and sustain unprecedented levels of stratospheric chlorine." [34].

Based on such advice, the Parties to the Montreal Protocol, meeting in London in 1990, agreed to phase out CFCs and halons by the year 2000; to extend the controls to any fully halogenated CFC (previously only named compounds were covered); to phase out carbon tetrachloride by 2000 and to phase out methyl chloroform by 2005. These controls extended to the developed world only. The ten year stay of execution for countries operating under Article 2 remained, as did an allowance of 10% or 15% (depending on the substance) that could be produced to meet the "basic domestic needs" of the Article 2 countries (see Sect. 2.2, below). In addition a mechanism was put in place to agree "essential uses" for which the materials could still be produced after other production had been phased out.

2.1.2
Copenhagen 1992

The London revisions to the Protocol had included a Resolution on Transitional Substances, laying down guidelines for the use of HCFCs, essentially promoting good working practices and calling for: "... replacement of the transitional substances by non-ozone depleting and more environmentally suitable alternatives... no later than 2040 and, if possible, no later than 2020." This suggestion came to be included in the Protocol when it was amended in Copenhagen in 1992 when HCFCs were included in a formula that set a "cap" on consumption and progressively reduced it to virtually zero by 2020, with complete phase-out in 2030. For each nation, the cap was set at the sum of its 1989 consumption of HCFCs plus 3.1 % of its total consumption of CFCs in that year. The cap percentage was the subject of intense negotiation and the result was a pragmatic compromise that reflected the findings of the Technical and Economic Assessment Panel [35], and the findings of scientific reports to the Parties [27] and governments [36].

The calculations for the cap are based on ODP tonnes (that is the mass of each substance consumed multiplied by its ozone depletion potential). This accommodates the lower intrinsic effects on the ozone layer of HCFCs which generally have shorter atmospheric lifetimes than CFCs as a consequence of their oxidation by hydroxyl radicals in the troposphere.

In addition to incorporating the transitional substances formally, the Copenhagen amendments brought forward the dates for phase-out of CFCs, carbon tetrachloride and methyl chloroform (all to 1996) and halons (to 1994). In part, this was in recognition of the far greater potency of bromine for ozone depletion than chlorine, as had been made clear in the WMO Scientific Assessment [27]. For the same reason, methyl bromide was formally included in the Protocol by this amendment with a freeze on consumption in the developed world in 1995.

2.1.3 *Vienna 1995*

In the interval between the Copenhagen and Vienna amendments, the first signs of the response of the environment to the Montreal Protocol could be discerned: the increase in concentrations of CFCs 11, 12 and 113 and of methyl chloroform had begun to slow down [37]. However, the major review of ozone depletion in 1994 [28] gave little grounds for complacency, particularly because the extent and severity of Antarctic ozone holes continued to increase in 1992 and 1993. In 1995, CFCs, carbon tetrachloride, methyl chloroform and halons were all but phased out in the developed world, so that there was scope for change only as regards transitional substances and methyl bromide. The cap percentage was reduced from 3.1 to 2.8% in this amendment and a phase-out schedule for methyl bromide was implemented. Both affected only the developed world.

2.1.4 *Montreal 1997*

By the time that the Parties met in Montreal, there was a clearly discernible response of the halogen loading of the atmosphere to the reductions in production and consumption of halocarbons, that actually had gone significantly faster than was required by the Protocol. Tropospheric chlorine-loading peaked in 1993, from which it could be inferred that maximum stratospheric chlorine concentrations would occur a few years later. The peak in bromine-loading could be expected to occur between the years 2000 and 2010 [38]. The Montreal amendments concentrated on consolidating the environmental improvements that had been made by the developed world and extending the controls on HCFCs and methyl bromide to the developing world. The current suite of controls is summarised in Table 1.

Table 1. Current controls under the Montreal Protocol

Industrial Countries		Article 5 (developing) Countries	
Substance and (baseline)	Schedule	Substance and (baseline)	Schedule
Principal CFCs, Annex A-I (1986)	July 1989: freeze 1994: −75% 1995: −100%	Principal CFCs, Annex A-I (average of 1995–1997)	July 1999: freeze 2005: −50% 2007: −85% 2010: −100%
Halons, Annex A-II (1986)	1992: freeze 1994: −100%	Halons, Annex A-II (average of 1995–1997)	2002: freeze 2005: −50% 2010: −100%
Other CFCs, Annex B-I (1989)	1993: −20% 1994: −75% 1996: −100%	Other CFCs, Annex B-I (1998–2000)	2003: −20% 2007: −85% 2010: −100%
Carbon tetrachloride Annex B-II (1989)	1995: −85% 1996: −100%	Carbon tetrachloride Annex B-II (1998-2000)	2005: −85% 2010: −100%
Methyl chloroform, Annex B-III (1989)	1993: freeze 1994: −50% 1996: −100%	Methyl chloroform, Annex B-III (1998-2000)	2003: freeze 2005: −30% 2010: −70% 2015: −100%
HCFCs, Annex C-I (1989, plus 2.8% of 1989 CFC consumption)	1996: freeze 2004: −35% 2010: −65% 2015: −90% 2020: −99.5% 2030: −100%	HCFCs, Annex C-I (2015)	2016: freeze 2040: −100%
Hydrobromo-fluorocarbons (HBFCs), Annex C-II	1996: −100%	Hydrobromo-fluorocarbons (HBFCs), Annex C-II	1996: −100%
Methyl bromide, Annex E (1991)	1995: freeze 1999: −25% 2001: −50% 2003: −70% 2005: −100%	Methyl bromide, Annex E (1995–1998)	2002: freeze 2005: −20% 2015: −100%

Sources [40–41]. Unless otherwise stated, dates refer to January 1st.

The latest reviews of the science and potential effects of ozone depletion and of the technologies for replacing ODS are taking place and will be published before the end of 1998 [39].

2.2 Adapting the Protocol to the Needs of Nations: Developing Countries

Most of the historic production and consumption of CFCs and other Ozone Depleting Substances (ODS) has been in the developed world [42, 43], in 1986 all but 4% of CFCs were produced there. Use of highly technical and expensive solvents, air-conditioning and even – to an extent – refrigeration, are symbolic of affluent economies. So, in the absence of restrictions, the use of CFCs could be expected to increase in the developing world as it industrialised. This was considered in the drafting of the Montreal Protocol, with the result that countries that have per capita consumption of CFCs less than 0.3 $kg \cdot year^{-1}$ and less than 0.2 $kg \cdot year^{-1}$ of halons are allowed to operate under Article 5 of the Protocol. This gives them flexibility to use ODS for longer than the developed world, in order to avoid hindering development but the Protocol retains ultimate control over the use of ODS, with agreed phase-out dates.

Although the allowance is written in terms of a ten year delay in compliance, the effective dates are those brought in with the London revisions. Consequently, Article 5 countries must phase out CFCs, Halons and carbon tetrachloride by 2010 and methyl chloroform by 2015. During subsequent revisions of the Protocol, the requirements for HCFCs and methyl bromide were laid down. For the former, consumption will be frozen in Article 5 countries in 2016 (at 2015 levels) until phase-out in 2040 and, for the latter, consumption will be frozen in 2002 at the average level during 1995 to 1998, with phase-out in 2015. The last provision, implemented during the Montreal revision, removed concern that continued expansion of use of methyl bromide in agriculture in developing countries would reverse the reductions in atmospheric halogen-loadings anticipated for the next century.

Between now and the phase-out dates in the developing world, the Article 2 countries may continue to produce ODS at up to 15% of the rate in 1986, in order to meet the basic domestic needs of the developing world. The quantity produced and the amount consumed in Article 5 countries are reported nationally to UNEP and, for CFCs and HCFCs, are entered by the companies in the AFEAS data collection. According to the latter, the total production of CFCs in 1996 was less than 8% of the 1986 level; this includes some production in Article 5 countries by subsidiaries of multinational companies [42].

Furthermore, Article 5 countries are permitted to operate domestic production facilities to meet their own, and other Article 5 countries' needs. Recognising that they might find the financial implications of phase-out onerous, the London Amendment provided for the creation of a Multilateral Fund. Its responsibilities were to meet the incremental costs of the Parties operating under Article 5 of the Protocol, to implement its control measures, to finance all clearing house functions: i.e., country studies, technical assistance, information, training and the costs of the Fund Secretariat. The Fund has so far disbursed about US $450 million and has been effective in closing down some CFC production facilities in the developing world and also in facilitating the transfer to non-ozone-depleting technologies for refrigeration, insulation and air-conditioning [44].

2.3 What Might Have Happened Without the Montreal Protocol

The early commercial history of CFCs and other Ozone Depleting Substances is characterised by rapid growth: in the range of applications, and in the geographical and societal extents of the applications, and – hence – in the quantities produced and consumed. This was partly driven by the chemical inertness of the substances and their low toxicity, two factors that made release into the atmosphere a normally accepted method for their disposal. In some cases, such as propellants for aerosol spray dispensers and solvents for adhesives, this release was an intrinsic part of the application; in others it was simply the most economical means of disposal of unwanted material. With the exception of some of the material used in insulating foam, most of the ODS were eventually released into the atmosphere.

In the free market that existed before 1974, CFCs showed remarkable growth. At that date, the combined annual production of CFC-11 and CFC-12 was more than 800 kt year^{-1} and had been growing at 10% year^{-1} for over two decades. Had the ozone depletion theory not been evinced by Molina and Rowland in 1974 [3], and had there not been a history of Antarctic ozone measurements dating back to 1956, that enabled the ozone hole to be identified as a recurrent phenomenon only a few years after the first spring in which significant depletion was observed [23], the first signs might have been severe, sudden changes to the ozone distribution in populated regions of the southern hemisphere. Without the warning sounded by Molina and Rowland and the comparatively early observations of Farman and his colleagues indicating that ozone was indeed being depleted, the necessity to construct instruments like the Vienna Convention and the Montreal Protocol would perhaps not have been recognized before 1993, with no history of previous negotiations [45]. Consequently, the 10% yr^{-1} growth in CFCs and other ODS might have been expected to continue, possibly for several decades. In reality, the growth was much more modest up to the first iteration of the Montreal Protocol; since then consumption has dropped at rates faster than those required by each of its revisions. From 1974 to 1990, total consumption of CFC-11 and CFC-12, the most widely used ODS, remained virtually constant and, – although other ODS such as CFC-113 continued to grow apace – they never reached the same levels.

In the situation of continued free-market growth, whether a modified continuation of the exponential growth (as suggested by Prather et al. [45] or an analytical approach such as the Gompertzian growth towards a saturation value (described in Properties and Applications of Industrial Halocarbons in this volume, Sect. 3) is adopted, the result would have been a massive increase in annual production, consumption and – by inference – emissions during the period up to 2002. Prather and co-workers suggest that, had the Antarctic ozone hole come as a surprise in the early 1990s, 2002 would have been be the earliest date at which a global ban on emission of CFCs could have been put in force. They estimated that, under those circumstances, the chlorine-loading of the troposphere would have reached 9 ppb (parts per billion, 9 in 10^9).

Chlorine loading is the total concentration of all chlorine-containing ODS, expressed as their chlorine content; calculated for the troposphere, it is a measure of the force acting to change stratospheric chlorine concentrations. Bromine loading of the troposphere gives similar information for potential bromine concentration changes in the stratosphere. Bromine and chlorine have very different potency for stratospheric ozone depletion; the former being between

40 and 100 times more effective than the latter. Furthermore, the effectiveness of individual ODS varies, due to the way in which each decomposes in the stratosphere and releases halogen there. Clearly, chlorine in a compound such as chlorodifluoromethane (HCFC-22), which decomposes by reaction with hydroxyl radicals – that are hardly present at all in the ozone layer, will have a much lower intrinsic effectiveness than chlorine in carbon tetrachloride, which undergoes nearly complete decomposition by photolysis in the stratosphere [46]. The concept of Equivalent Effective Stratospheric Chlorine was devised to reconcile the differences between the potencies of ozone depleting substances, and to provide a mechanism to gauge the effectiveness of control measures that went a step beyond the simple halogen-loading calculations that had been used [34, 36, 37, 47]. The calculation involves adjusting chlorine and bromine loadings from individual substances according to the effectiveness factors described, for example, by Daniel et al. [46], and then displacing the adjusted concentration by a time-lag of three years, to allow for the delay in transporting the substance into the stratospheric ozone layer from the troposphere. This time-lag is based, in part, on tracer studies [48]; the calculations are described in detail in the later Scientific Assessments [28, 39] and Table 2 gives atmospheric lifetimes and adjustment (effectiveness) factors, for the principal ODS controlled under the Montreal Protocol.

Expressed as Equivalent Effective Stratospheric Chlorine (EESC), the 9 ppb of tropospheric chlorine estimated by Prather et al. [45] would be about 7 ppb, and should occur in 2005. In fact, EESC reached a very much lower peak value in 1996

Table 2. Atmospheric lifetimes and the effectiveness factors for halogen in ozone depleting substances controlled under the Montreal Protocol

Compound		Atmospheric Liftime (years)[a]	Halogen Effectiveness Factor[b]	
CFC-11	CCl_3F	50		0.91
CFC-12	CCl_2F_2	102		0.55
CFC-113	CCl_2FCClF_2	85		0.68
Carbon tetrachloride	CCl_4	42		0.96
Methyl chloroform[c]	CH_3CCl_3	4.8		0.98
Halon-1301	$CBrF_3$	65		29
Halon-1211	$CBrClF_2$	20	chlorine	1.00
			bromine	40
Methyl bromide[d]	CH_3Br	0.7		39
HCFC-22	$CHClF_2$	13.3		0.32
HCFC-141b	CH_3CCl_2F	9.4		0.65
HCFC-142b	CH_3CClF_2	19.5		0.33
HCFC-123	$CHCl_2CF_3$	1.4		1.01
HCFC-124	$CHClFCF_3$	5.9		0.47

[a] from [28] unless otherwise stated.
[b] from [46].
[c] lifetime from [32].
[d] lifetime from [39].

and it has since started to decline slowly. The rise in EESC, calculated from atmospheric measurements in the late 1960s and again in 1977 and 1990, is shown in Fig. 1 as a cumulative plot for the principal ODS. These are: methyl chloride (which is natural in origin), methyl bromide (both natural and anthropogenic), CFCs, methyl chloroform, carbon tetrachloride, HCFCs and halons. For the period 1990 to the present, the EESCs in Fig. 1 were calculated for each set of compounds from their atmospheric concentrations, measured year by year. For the future, the EESCs were calculated from the decay in current atmospheric concentrations of the substances, augmented by emissions from the "banks" of ODS still being used in equipment. These banks were themselves calculated from production records, as described in the AFEAS and UNEP compendia [42, 43], and augmented by the continuing productions permitted by the Montreal Protocol. With the exception of methyl chloroform and halons, production of which in the developed world has ceased [43], the scenario uses the full allowances of the Protocol and can thus be considered to be a realistic maximum. In the past, the reductions in ODS production and consumption have been greater and have occurred sooner than required under the Protocol.

On the basis of this scenario, stratospheric halogen will return by the early 2050s to the levels which existed in the late 1970s, when the annual Antarctic ozone hole first became discernible. If all emissions were to cease forthwith, the

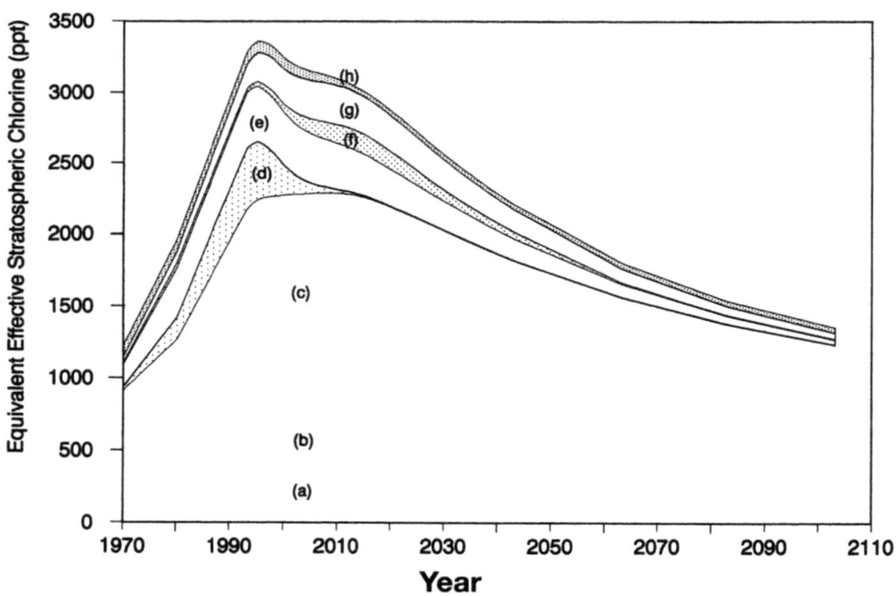

Fig. 1. Past, present and future Equivalent Effective Stratospheric Chlorine on a maximum scenario with full compliance with the Montreal Protocol. Blocks (*a*) and (*b*) are principally from natural methyl chloride and methyl bromide, respectively, (*c*) from all CFCs, (*d*) from methyl chloroform, (*e*) from carbon tetrachloride, (*f*) from all HCFCs, (*g*) from all halons and (*h*) from that part of methyl bromide use that is potentially controllable under the Protocol

recovery would proceed at a pace dictated by removal processes in the atmosphere and the return to such levels of EESC would occur 20 years earlier – in about 2030. This is the earliest that ozone loss in either the Antarctic or mid-latitudes can be expected to cease, assuming no interference from other factors, such as stratospheric temperature [39]. By contrast, if Prather and co-workers' scenario had materialised, and the Montreal Protocol been delayed less than ten years, ozone losses would have been more severe and have persisted well into the 22nd century [45].

3
Climate Change and the Kyoto Protocol

This is arguably the next great environmental challenge to governments. The way that the threat of climate change from the accumulation of greenhouse gases has been addressed by international regulations bears some similarity to the negotiation of the Montreal Protocol and the scientific assessment of the two processes share a common heritage.

The concept that atmospheric gases which absorb infra-red radiation would affect the climate was suggested at the beginning of this century [49]. However, many years elapsed before the proposition was subjected to detailed examination, when the climatic implications of increasing concentrations of greenhouse gases were included in some of the earlier assessments conducted under CCOL and the Vienna Convention [11, 24]. These examined the physics of the atmospheric effects of increasing greenhouse gases and ozone depletion, but the first scientific reports that addressed all the implications, from the dynamics and possible detection of climate change through to its potential impacts on society, were those of the Intergovernmental Panel on Climate Change [50, 51]. These reports provided the scientific bases for the negotiations that commenced in February 1991 with the first meeting of the UN's International Negotiating Committee and resulted in the United Nations Framework Convention on Climate Change (the Rio Convention). This has the ultimate objective of stabilisation of greenhouse gas concentrations in the atmosphere at a level that would prevent dangerous anthropogenic interference with the climate system. It goes on to recommend that "Such a level should be achieved within a time-frame sufficient to allow ecosystems to adapt naturally to climate change, to ensure that food production is not threatened and to enable economic development in a sustainable manner". However, in keeping with the nature of a Framework Convention, the actual provisions are somewhat less draconian, and the Parties to the Convention were required to adopt policies and take measures that are simply aimed at limiting anthropogenic emissions of greenhouse gases. The Rio Convention bears the same relationship to climate change as the Vienna Convention to ozone depletion; similarly, the more rigorous controls are contained in Protocols to the Convention, the first of which is the Kyoto Protocol [52].

In order for a gas to be implicated in climate change, it must both absorb infra-red radiation and accumulate in the atmosphere. The first of these attributes is a molecular property, so the extent to which a gas will affect atmospheric temperature radiatively can be calculated relatively simply from its infra-red absorption spectrum and a model of the natural transmittance of infra-red radiation through the atmosphere, as described by Ramanathan [24]. The second is a consequence of imbalance between the rate of addition of the material to the atmosphere – the source flux – and its rate of removal, that its related to its atmospheric lifetime. Gases with long lifetimes, such as perfluorocarbons (PFCs) can accumulate in the atmosphere even if their fluxes are relatively small [53]. At the other extreme, a gas that has a short lifetime can accumulate to radiatively important concentrations, provided that its flux is large enough. This is the case for tropospheric ozone that has a lifetime of a few weeks at the earth's surface, but accounts for about 15% of the calculated climate forcing, due to the very large "secondary" flux arising from atmospheric reactions of hydrocarbons and oxides of nitrogen [54].

The most important primary atmospheric greenhouse gas is carbon dioxide (CO_2), which accounts for 64% of the increase in radiative forcing since pre-industrial times. Methane (CH_4) and nitrous oxide (N_2O), together, are calculated to contribute 28% and halocarbons the remaining 6%. The halocarbon contribution is expected to fall to 1.5% by the year 2050, partly due to the fall in flux and atmospheric concentration resulting from the controls of the Montreal Protocol; but the rise in CO_2 concentration depresses the relative contribution of other gases significantly. Figure 2 shows the expected contribution to future radiative forcing from the primary greenhouse gases. Secondary greenhouse gases, such as ozone and water vapour, are not included, on the grounds that their effects vary geographically and were too difficult to quantify with the same accuracy as those of the primary gases [55].

Carbon dioxide is, intrinsically, not a particularly powerful greenhouse gas. Nevertheless it has a very long environmental lifetime, so that the influence of an emission of CO_2 persists for many hundreds of years. Figure 3 shows how the absolute global-warming potential of CO_2 tends slowly towards a constant value over a timescale measured in centuries. Furthermore, there is a massive annual release of CO_2 into the environment, 26,000 million $t \cdot year^{-1}$ coming mainly from fossil fuel combustion; hence its large contribution to global warming and climate change.

Because of its position as the pre-eminent greenhouse gas, CO_2 is the reference compound against which the intrinsic effects of other greenhouse gases are judged, expressed as the ratio of the radiative forcing effect of a release of one kilogram of the target compound to the effect of a kilogram of CO_2. The problem that the effect of CO_2 changes with time has been addressed by integrating its radiative forcing effect, as well as that of the other greenhouse gases, only up to a particular year (the Integration Time Horizon). The effect of this is to include progressively more of the effect from CO_2 as the time horizon lengthens, so that – as a general rule – GWPs decrease with longer time horizons. For most purposes, a time horizon of 100 years is used, but GWPs are also reported at 20 year and 500 year time horizons.

Halocarbons are effective absorbers of infra-red radiation, so that their GWPs are appreciable, in the range 4 to 12,000 (at 100 year ITH) [58];(see also Table 1 in Properties and Applications of Industrial Halocarbons, this volume). Consequently, halocarbons in the forms of hydrofluorocarbons

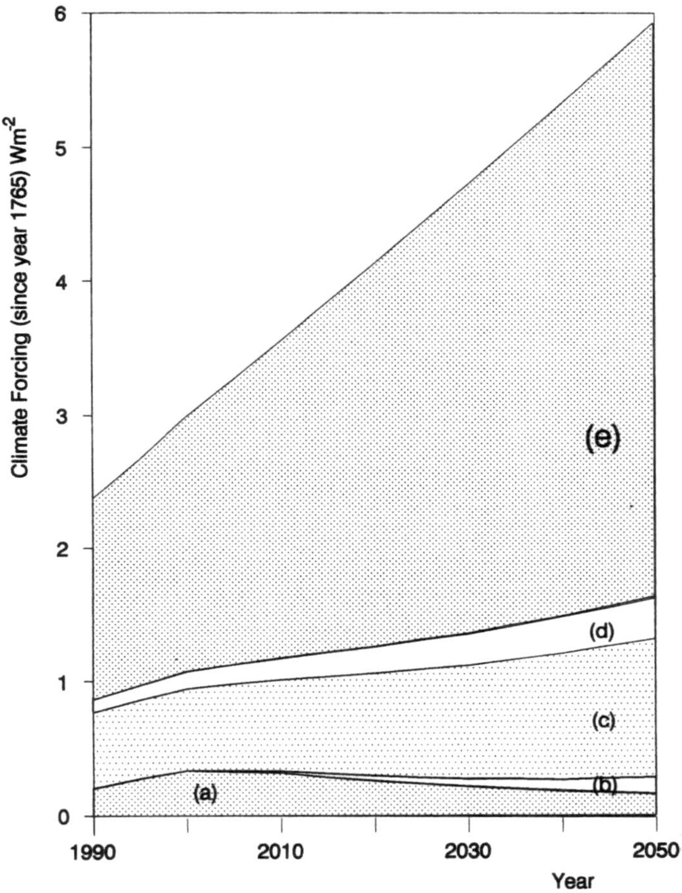

Fig. 2. Calculated Future Radiative Forcing from Primary Greenhouse Gases emitted at rates envisaged in the Reference Scenario (IS92a) from IPCC. [56–58]. Block (*a*) is the contribution from the gases controlled under the Montreal Protocol (CFCs and HCFCs), (*b*) is the contribution from future HFCs, (*c*) from methane, (*d*) from nitrous oxide and (*e*) from carbon dioxide

(HFCs) and perfluorocarbons (PFCs) have been included in the Kyoto Protocol as part of the "basket" of greenhouse gases, emissions of which must be reduced. The other gases included are CO_2, CH_4, N_2O, and sulphur hexafluoride (SF_6) [52].

A significant commitment under the Rio Convention was the provision of inventories of national emissions of greenhouse gases and, for the halocarbons involved, a tiered methodology was evolved by OECD (Organisation for Economic Cooperation and Development) and IPCC, working together. The objective was to provide reference methodologies against which national accounting could be standardised [59]. At the very simplest, the emissions in any year are set

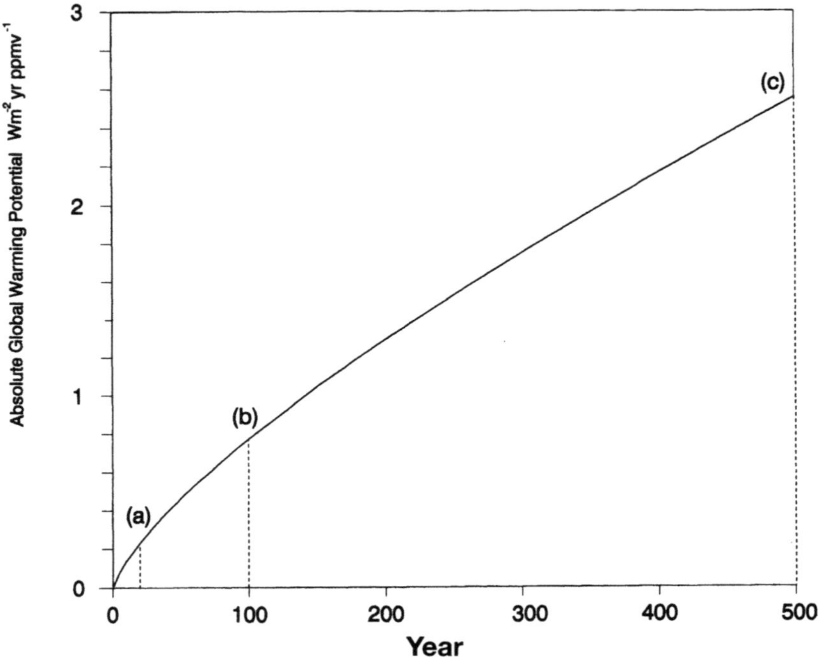

Fig. 3. The Evolution in Time of the Absolute Global Warming Potential (AGWP) of Carbon Dioxide. The AGWP at time horizons of (*a*) 20, (*b*) 100 and (*c*) 500 years provides the reference for the Relative GWPs of other gases

equal to the consumption (i.e. production plus bulk imports less bulk exports) in that year. This has the effect of accelerating emissions of the gases unrealistically, but it sets a maximum. At the other end of the scale of complexity, the actual emissions may be estimated from a detailed, item-by-item analysis of emission rates, quantities in use, and disposal procedures in each application. Nations are free to use their own methods and factors, but they must be able to justify changes from the OECD/IPCC methodology. Secondary greenhouse gases, such as non-methane hydrocarbons and oxides of nitrogen, that can generate tropospheric ozone, are also included in the methodology of the emissions inventory.

The standard emissions inventories will achieve more importance due to the Kyoto Protocol to the UNFCCC which sets targets for reductions in greenhouse gas emissions. Using 1990 emissions as the baseline, the "aggregate anthropogenic carbon dioxide equivalent emissions" of the greenhouse gases described above must be reduced overall by at least 5% in the period 2008 to 2012. Carbon dioxide equivalence is actually the mass of the emission multiplied by the 100 year Global Warming Potential of the gas concerned. Targets are, in fact, variable. The nations of the European Union, for example, have targets within the Kyoto Protocol of 8%, while the target for the USA is 7% and some nations are allowed to increase releases of greenhouse gases – notably Australia, which is allowed an 8% increase. In recognition of the fact that, in 1990, emissions of the halocarbon greenhouse gases not controlled by the Montreal Protocol were very small, 1995 can be used as the base year for HFCs, PFCs and SF_6 [52].

4 In Conclusion

The Montreal Protocol is beginning to have the desired effect, enabling the scientists conducting the latest assessment to conclude "The total combined abundance of ozone-depleting compounds in the lower atmosphere peaked in about 1994 and is now slowly declining." and "The combined abundance of stratospheric chlorine and bromine is expected to peak before the year 2000." and that, although "unambiguous detection of the beginning of the recovery of the ozone layer is expected to be well after the maximum loading of ozone depleting gases", "The rate of decline in stratospheric ozone at midlatitudes has slowed" [39].

Although there is superficial similarity between the politics of ozone depletion and those of climate change, and indeed much scientific interaction between the two, climate change has much wider implications. The range of materials and activities to be considered in regulations and the range of consequences are far larger for climate change and, because of the very long lifetime of carbon dioxide, the timescale for recovery from any effect on climate is far longer. Nevertheless, the Kyoto Protocol is an important first step.

5 References

1. Lovelock JE (1971) Nature 230:379
2. Lovelock JE, Maggs RJ, Wade RJ (1973) Nature 241:194
3. Molina MJ, Rowland FS (1974) Nature 249:810
4. Rowland FS, Molina MJ (1975) Rev Geophys and Space Phys 13:1
5. Reinsel G, Tiao GC, Wang MN, Lewis R, Nychka D (1981) Atmos Environ 15:1569
6. St. John DS, Bailey SP, Fellner WH, Minor JM, Snee RD (1981) J Geophys Res 86:7299
7. NAS (1979) Stratospheric ozone depletion by halocarbons: chemistry and transport. NAS, Washington
8. Derwent RG, Curtis AR (1981) Stratospheric ozone depletion estimates for global halocarbon usage estimated by the linear superposition of contributions from individual halocarbons, Report No. AERE-R 10168. HMSO, London
9. NASA (1979) The stratosphere: present and future, reference publication no 1049. NASA, Washington
10. DoE (1979), Chlorofluorocarbons and Their Effect on Stratospheric Ozone. HMSO, London
11. WMO (1981) The stratosphere 1981: theory and measurement. WMO, Geneva
12. NRC (1982) Causes and effects of stratospheric ozone reduction: an update. National Academy Press, Washington
13. McCarthy RC, Bower FA, Jesson JP (1977) Atmos Environ 11:491
14. Jesson JP (1980) Release of industrial halocarbons and tropospheric budget. In: Proc NATO Adv Study on Atmos Ozone. US Fed Aviat Administration, Washington, p 373

15. Fisher DA, Midgley PM (1993) Atmos Environ 27 A:271
16. McCulloch A (1992) Atmos Environ 26 A:1325
17. Neely WB, Plonka JH (1978) Env Sci Technol 12:317
18. Midgley PM (1989) Atmos Environ 23:2663
19. Jesson JP (1977) Angew Chem Int Ed Engl 16:513
20. OJ (1978) Council resolution of 30 May 1978 on fluorocarbons in the environment. OJ (C) 133:1
21. OJ (1980) 82/372/EEC: Council decision of 26 March 1980 concerning chlorofluorocarbons in the environment, OJ (L) 090:45
22. UNEP (1998) Status of ratification/accession/acceptance/approval of the agreements on the protection of the stratospheric ozone layer. UNEP, Geneva
23. Farman JC, Gardiner BG, Shanklin JD (1985) Nature 315:207
24. WMO (1985) Atmospheric ozone: 1985, assessment of our understanding of the processes controlling its present distribution and change. WMO, Geneva
25. WMO (1988) Report of the international ozone trends panel. WMO, Geneva
26. WMO (1989) Scientific assessment of stratospheric ozone: 1989. WMO, Geneva
27. WMO (1991) Scientific assessment of ozone depletion: 1991. WMO, Geneva
28. WMO (1994) Scientific assessment of ozone depletion: 1994. WMO, Geneva
29. TEAP (1994) 1994 Report of the technology and economic assessment panel and its technical options committees. UNEP, Nairobi
30. UNEP (1994) Environmental effects of ozone depletion: 1994 assessment. UNEP, Nairobi
31. SORG (1988) Stratospheric ozone 1988, second report of the UKSORG. HMSO, London
32. Prinn RG, Weiss RF, Miller BR, Huang J, Alyea FN, Cunnold DM, Fraser PJ, Hartley DE, Simmonds PG (1995) Science 269:187
33. Midgley PM, Fisher DA (1993) Atmos Environ 27 A:2215
34. SORG (1990) Stratospheric ozone 1990, third report of the UKSORG. HMSO, London
35. TEAP (1991) 1991 Report of the technology and economic assessment panel to the parties to the montreal protocol. UNEP, Nairobi
36. SORG (1991) Stratospheric ozone 1991, fourth report of the UKSORG. HMSO, London
37. SORG (1993) Stratospheric ozone 1993, fifth report of the UKSORG. HMSO, London
38. SORG (1996) Stratospheric ozone 1996, sixth report of the UKSORG. UK Department of the Environment, Report No. 96DPL0021, London
39. WMO (1998) Scientific assessment of ozone depletion: 1998. WMO, Geneva
40. UNEP (1997) The 1987 Montreal protocol on substances that deplete the ozone layer as adjusted and amended by the second meeting of the parties (London, 27–June 1990) and by the fourth meeting of the parties (Copenhagen, 23–November 1992) and further adjusted by the seventh meeting of the parties (Vienna, 5–December 1995) and further adjusted and amended by the ninth meeting of the parties (Montreal, 15–September). UNEP, Nairobi
41. Oberthür S (1997) Production and consumption of ozone-depleting substances. Deutsche Gesellschaft für Technische Zusammenarbeit, Eschborn
42. AFEAS (1998) Production, sales and atmospheric release of fluorocarbons through 1996. AFEAS, Washington
43. UNEP (1997) Production and consumption of ozone depleting substances, 1986–1995. UNEP, Nairobi
44. UNEP (1997) Report of the chairman of the executive committee of the multilateral fund, in report of the ninth meeting of the parties to the montreal protocol on substances that deplete the ozone layer, document no. UNEP/OzL.Pro.9/12. UNEP, Nairobi
45. Prather MJ, Midgley PM, Rowland FS, Stolarski R (1996) Nature 381:551
46. Daniel JS, Solomon S, Albritton DL (1995) J Geophys Res 100:1271
47. Prather MJ, Watson RT (1990) Nature 344:729
48. Pollock WH, Heidt LE, Lueb RA, Vedder JF, Mills MJ, Solomon S (1992) J Geophys Res 97:12,993
49. Arrhenius S (1909) Centr Min: 481
50. IPCC (1990) Houghton JT, Jenkins GJ, Ephraums JJ (eds) Climate change: the IPCC scientific assessment. University Press, Cambridge

51. IPCC (1990) Bernthal FM (ed) Climate change: the IPCC response strategies. World Meteorological Organisation, Geneva
52. UN (1977) Kyoto protocol to the Framework Convention on Climate Change, UN document ref FCCC/CP/1997/L.7/Add.1. UN, Bonn
53. Harnisch J, Borchers R, Fabian P, Maiss M (1996) Geophys Res Lett 23:1099
54. van Dorland R, Dentener FJ, Lelieveld J (1997) J Geophys Res 102:28,079
55. IPCC (1994) Houghton JT, Meira Filho LG, Bruce J, Lee H, Callander BA, Haites E, Harris N, Maskell K (eds) Climate change 1994: Radiative forcing of climate change and an evaluation of the IPCC IS92 emission scenarios. University Press, Cambridge
56. Wigley TML, Holt T, Raper SCB (1991) STUGE (an Interactive Greenhouse Model): User's Manual, Climate Research Unit, University of East Anglia, Norwich
57. IPCC (1992), Houghton JT, Callander BA, Varney SK (eds) Climate change 1992: the supplementary report to the IPCC scientific assessment. University Press, Cambridge
58. IPCC (1996) Houghton JT, Meira Filho LG, Callander BA, Harris N, Kattenberg A, Maskell K (eds) Climate change 1995: the science of climate change, contribution of Working Group 1 to the second assessment report of the IPCC. University Press, Cambridge
59. IPCC (1995) IPCC guidelines for national greenhouse gas inventories. IPCC WG1 Technical Support Unit, Bracknell

Subject Index

The Handbook of **Environmental Chemistry**

Volume 5 PT C

J. Hřubec (Ed.)

Quality and Treatment of Drinking Water II

1998. XV, 180 pp.
Hardcover DM 198,-
ISBN 3-540-62574-7

Drinking water quality is a vast and complex subject. In addition to the topics already addressed in Volume 5, part B, this new volume discusses in an authoritative way the current key issues of drinking water quality and its control: - Toxicity tests for assessing drinking water quality - Toxicological approaches for developing drinking water standards - Analysis of organic micropollutants - Algal toxins and human health - Quality changes due to application of ozone and chlorine dioxide.
The articles are written by leading experts and present the state of the art of drinking water research. This volume will therefore be a valuable source not only for scientists and engineers, but also for decision-makers in government, environmental control and industry.

MIX
Papier aus verantwortungsvollen Quellen
Paper from responsible sources
FSC® C105338

If you have any concerns about our products,
you can contact us on
ProductSafety@springernature.com

In case Publisher is established outside the EU,
the EU authorized representative is:
Springer Nature Customer Service Center GmbH
Europaplatz 3, 69115 Heidelberg, Germany

Printed by Libri Plureos GmbH
in Hamburg, Germany